AT HOME WITH ALTERNATIVE ENERGY

A Comprehensive Guide to Creating Your Own Systems

BY MICHAEL HACKLEMAN

PEACE PRESS

© 1980 Michael Hackleman
All rights reserved.

Peace Press, Inc.
3828 Willat Avenue
Culver City, California 90230

9 8 7 6 5 4 3 2 1

Typesetting by Freedmen's Organization,
Los Angeles, California
Printed in the United States of America by Peace Press

Library of Congress Cataloging in Publication Data

Hackleman, Michael A
    At home with alternative energy.
    Includes index.
    1.   Renewable energy sources.   I. Title.
TJ163.2.H3   621.4   79-48056
ISBN 0-915238-38-1

Contents

Foreword 1

1. **ENERGY CROSSROADS 3**
   Changes 3

2. **SOLAR ENERGY 5**
   Collector Types 9
   Medium Transfer 19
   Storage 25
   Passive Space-heating Systems 27
   Getting the Rays 31
   Collector Characteristics 36
   Codes on Solar 43
   Backup Heating 45
   Specific Applications 45
   Cooling Techniques 52
   Final Solar Checklist 56

3. **WIND ENERGY 60**
   Wind Characteristics 60
   How Much Energy 61
   Aeroturbines 65
   Windplant Ratings 74
   Storage Systems 75
   Automatic Windplant Control 78
   Using Wind Electricity 83
   Towers 86
   Wind for Pumping Water 89
   Windplant Costs 92
   Final Notes 94

4. **WOOD, WATER, AND METHANE 95**
   Wood Energy 95
   Water Power 102
   Methane Power 116

5. **INTEGRATION 124**
   Refrigeration 124
   Cooking 125
   A Standby Generator 125
   Final Words 125

APPENDIX: DATA CUBBYHOLE 126
   A. Sizing Solar Energy Storage 126
   B. Angle of Declination Calculations 127
   C. Aeroturbine Efficiency and the Windpower Formula 128
   D. Aeroturbine Efficiencies 129
   E. Windpower Formula Calculations 129
   F. Estimating Wind at Varying Heights Above Ground Level 129
   G. Flow Rate Calculations Methods for Small Streams 130
   H. Measuring the Head of Water 134
   I. An Alternate Gearing Technique for Waterwheels 135
   J. Conversion Table 136

SOURCES AND REFERENCES 137

INDEX 142

Foreword

One writes to say something. How well it's said is a matter of personal opinion and taste, and depends a lot on how interested one is in the subject. Sometimes an author writes for himself or herself, impervious to and uncaring of whoever else might wade through the completed text. Others write for a specific audience; in the writer's mind is an image of the reader of this material. So it was with me. Before each sitting, I conjured up an audience. So that you won't get into this book mistakenly, I'll tell you whom I saw.

My readers are tired of "how-to" books, wanting a "why-to" book that places them at home with the principles inherent in any design. These individuals don't want another "plan"-type book; rather, they feel the need for a strategy to help evaluate the multitude of plans now available. My audience is not expecting something for nothing, knowing that it's hard enough to get what you pay for, much less anything more. They speak in action, doing what others talk about, year after year. These readers cannot afford an architect—only the paper, pencil, and patience to design their own home. They are people who have discovered the fallacy of buying and owning their own land but still "renting" energy. They have learned that, with the best of intentions, a system can be "mismatched" to an application through ignorance or misunderstanding between seller and buyer. My readers want to know which factors are essential to an alternative energy system, which are a matter of choice, and which occurs when.

I don't have exclusive rights to this audience. But even with the same group of people as readers, each writer will choose a different path. Some books are packed full of theory and equations, ignoring the value of simplicity and easy reading. Other books deal with specific plans for a range of workable units; I have a library full of them, and use them often. However, I find myself modifying these designs to fit my own circumstances and, in doing so, sometimes I irritatingly uncover a factor which cannot be compromised. And, of course, there's the "industrial survey" book which covers the potential of energy in the future in terms of the cost per ft^2 per Btu per high-intensive, technological material per unit person, which I might read if I didn't have a Yellow Pages handy.

This book is intended to bridge the field, plug a few gaps, and to be adaptable to whatever works for you—your locality, finances, talents, skills, obsessions, and needs. Even if you don't envision yourself with a hammer in hand, you will need to arm yourself in the fight to buy a general design that will fit your situation. I think there's much joy and pride to be had from doing anything yourself, but there are limits, and you *must* enjoy it to some extent or it becomes a liability. Either way, the more you know, the less you pay. You may not want to deal with designing your own system, but the way I see it, it's better to know how and decide not to than to want to and not know how.

In this book, we'll be dealing with natural processes. Rather than starting someplace and ending up somewhere else in a very linear way, they go in circles, one leading into the other. Everything follows and everything precedes. A perfect description, incidentally, of the planet; it has much to tell us. One is that you don't put all your cherries in one pie. That is, no single source of energy does it all, all of the time. Another is that while the source is free, the initial investment is certainly not. TANSTAAFL—There Ain't No Such Thing As A Free Lunch.

I'll tell you right off that I'm not going to try to convince you to use alternative energy. If you don't like the idea, by all means build a nuclear power plant. I also don't make assumptions about what you do and don't know. If you just squeaked by in math or science in college, your secret is safe with me; prior knowledge in these areas isn't required.

A word about metrics: When this book was first

written, all measurements of dimension, weight, volume, velocity, temperature, work, power, etc., were given in the U.S. Customary System of measurement —inches, pounds, square feet, etc.—with the metric equivalent in parentheses. The most noticeable effect of doing this was a reduction in readability of the text. In many cases, this also bogged down discussions where the actual figures used were of no particular value in and of themselves, but only examples to give the reader a feel for the subject. Giving equivalents in some instances and not in others didn't seem the best way to proceed either. Finally, the author and the publisher agreed that a metric conversion table (see Appendix: Data Cubbyhole, Section J) solved the problem in the neatest way; I'm sure the metric advocate won't agree, but that's what we've done.

There is stout resistance to metrification in all countries now using the U.S. system, but, without going into the multitude of advantages and disadvantages to either system or this changeover, all books will suffer somewhat in the interim. Since this book is dealing with ideas and *not* meant as a technical treatment of the subjects discussed, I hope the metric convert will not find it too difficult to get past all the *numbers*.

To say any more would be to steal from the book's contents, so here's where you make your first decision: read on or put it back on the shelf!

I measure my life by the energy of my friends: our good friend Harold Moskovitz; a generous contribution of photos from a friend and one of the authors of *Living with Energy*, Ronald Alves; determination in a fellow worker, Bill Wilson; inspiration from John Mudie; imagination from Ed O'Brian; relief with David and Ray at the local chess pit; vehicle bail-outs from John and Trish; conspiracy from Gary, Barry, and Kent; hot water from Jim DeKorne and tower stubs from Windy; music from Bob and Doi and laughter from Bob Kaminski's puppet shows; competence from our typist, Diane; and love, understanding, tolerance, and support from my life's mate, Nessie. I dedicate this book to my future helper, Brett, who is presently content to laugh at all our insane antics.

1. Energy Crossroads

It may not always seem like it, but we live in a great age—unlimited energy and unlimited potential. Transportation is at the turn of a key, heat at the turn of a thermostat, and light and power at the flip of a switch. A lot of that comes from sources of energy presently on tap. We can count them on the fingers of just one hand: oil, gas, coal, hydroelectricity, and nuclear power. The Big Five.

Other sources of energy are being tapped to bring us minute amounts of energy now and, according to the utilities, a lot more in the future. We're now going after the high-sulfur oils (tar sands and shale), geothermal (earth heat) and hydrothermal (ocean cool) energy, tidal and wave energy, fusion and biofuels, and orbiting satellites that beam down the power of space-caught sunlight.

A fight has been brewing for years. Labeled malcontents, some folks think that there's something dreadfully wrong with the way energy for the future looks. Or the way the future looks, period. And while it used to be longhairs and eccentrics and radicals, a lot of pretty straight-looking people are getting involved. Just what is it they're talking about?

Alternatives. Different sources of energy from the ones we're now using and plan to use in the future. What alternatives? Briefly, they're sun, wind, water, wood, and methane. Others exist, such as tidal and wave energy and biofuels. But the first five are the most promising. What can the Alternative Five do?

Solar energy is everywhere abundant on planet Earth. In truth, it is the source of *all* energy on this planet except the uranium for nuclear power and geothermal energy. All fossil fules are just that—fuel from fossils. And no matter how you see it, that's stored-up sun energy. Earth is a planet-sized battery for solar power. And we're hung up on the energy from the battery (oil) and ignoring the power source that charged it in the first place! We can use the sun's energy directly as heat, or convert it to electricity. If the sun were to die, so would this planet and everything on it. Afterward, nuclear power would be only a short-lived flashlight in the dark.

Wind energy comes from solar energy. No kidding. It is the uneven heating of the earth's surface that causes air to rush about, trying to establish order and equilibrium. Wind is variable but somewhat predictable, and we can use funny-looking machines to extract some of its energy.

Wood energy is derived from living plants (hard to think of a tree as a plant, isn't it?), which depend heavily upon the sun's energy to grow in the first place. While wood is the source of many products and applications, it can be burned to produce heat, and, carefully managed, this is a renewable resource.

Water energy is already tapped for generating electricity on a large scale, but a multitude of streams and rivers wind their way across private lands where, with skill and understanding, individuals can install their own power sources. Water energy comes from solar energy, as well; the rain that falls on the mountains is the end of a "river" (composed of solar-evaporated ocean water) that flows in the sky.

And *methane* energy comes from the decomposition of organic things—plants, animals, and excrement—in a man-made version of the swamp. The resultant biogas is not unlike the petroleum gases we use today as propane, butane, and natural gas, and may be used as a substitute for them.

CHANGES

It's past time to make some decisions. Some folks think that if we don't start making them soon, it won't be long before future ones will be made for us. It's time to bone up on the problems associated with today's Big Five sources of energy (see the Sources and References section).

Our next step is to find a means of evaluating any source of energy fairly. What are the criteria?

I've got a few suggestions. The ideal source is renewable, clean, safe, practical, cheap, and readily abundant (for individual- and utility-size use). Make up your own list, delete any that I've presented that aren't of concern to you, add any of your own, and take 'em one by one, giving a point for each criteria satisfied by each source. In my own evaluation, the final tally came to 14 points for the Big Five and over 33 points for the Alternative Five.

Few people doubt the sanity of using alternative sources of energy. The discord comes when implementation is discussed or the implications of transition are contemplated. It seems we can't make the change without unacceptable sacrifices or virtual collapse of the nation's economy, if we're to judge the opposition's viewpoint correctly. Furthermore, there's the embarrassing answer to why there's so little interest by the utility companies in the alternatives: it's simply that they can't sell or control what is abundant everywhere and readily available to all. How does one go about shutting off the sun or wind when someone doesn't pay the utility bill?

What can you do? First of all, learn something about the subject. (Easily accomplished by reading the rest of this book!) Second, find out which of these sources are available to you. Third, decide whether you are willing to expend the effort in money, time, patience, and personal energy. Finally, stop voting! I'm not referring to the election kind of vote. This culture runs on economics. Paying the utility bill is a vote for whatever policy or energy source the utility wishes to use. If you don't like what they're doing, don't give them your vote. Simply not paying your bill will, of course, have them at your doorstep. Avoid this trouble by establishing your alternative energy source and disconnecting. If sanitation codes prohibit disconnecting, minimize your usage or simply throw your habitat's main breaker to "Off" as frequently as possible, as you do when going on vacation. If you can't prevent the vote, you can at least minimize it.

This doesn't have to be permanent. Try it for a weekend first—no TV, transportation, or lights. Go to bed early, get up early, be with family and friends. Then, try a week. Just shut off the main breaker. Then, a month. It doesn't have to be cold turkey, but it's a nice feeling to be off the utility fix. Get some friends in on your good time. A stay-at-home demonstration. Play some volleyball or Frisbee. Read the rest of this book and learn how you can make it a permanent vacation. Do it better yourself. Have your own blackouts if your homebuilt system goofs up. Get cold once in a while; it makes you appreciate warmth. It's all part of the process of exploring alternatives, whether they be energy or life-styles. One follows the other. Good hunting.

Fig. 1-1. *(Courtesy Earle Rich.)*

2. Solar Energy

Before we immerse ourselves in the particulars of solar energy systems, it would be a wise step to consider to what purpose the collected energy will be put, if for no other reason than to enable us to answer questions about our project.

Ask a hundred people what solar energy can be used for, and you'll get eighty-four who answer: *heating water*. Obviously, most of these people will have tried to get a quick, cool drink from a hose that's been sitting out in the summer sun. About three persons out of this hypothetical hundred will answer: *heating a house*. They'll be quick to add, "during the winter," immediately revealing that they live in upstate New York. A glasses-pipe-and-briefcase type will surely answer: *producing electricity*. Adding, "by photovoltaic means, of course"—to you and me, that's solar cells. Two mothers with very green thumbs will give us: *dehydrating food*. Along with canning, this is an excellent means of liberating energy-consuming freezers to deal with more perishable items. Coming in on the tail end of that answer, will be a gentleman who assures us, despite the apparent contradiction, that there is such a thing as a *solar refrigerator*. And, although he considers it *very* decadent, a *solar air conditioner*. A shocked observer will suggest a less expensive solar cooling technique: *solar shading*. Another passerby will stop, look over our list, and ponder the possibilities, finally suggesting (with a reservation for midnight snacks) *solar cooking*. A shriveled desert rat will wander by and proclaim the simplicity of: *solar distillation*. Clearing his throat, he will avoid the query forming on my lips by adding, "getting pure water from brackish or salt water," and ramble off. A few young boys will stop by to add their own answer: *Steam generation*. Well, to be honest, they might indicate the heating of any semi-volatile fluid for the purpose of gas expansion for mechanical or electrical energy production, but it sounds like steam generating to me. A young college student will wait till no one is in sight and then hastily explain the merits of her idea: *solar welding*.

And there you have it. Actually it might have been less involved to list what solar energy *can't* do. If you've taken a count of the number of persons responding to our question, you'll notice that we're shy three. What would they have to say? Well, one might launch into quite a dissertation on light and tones and body health. Another would join in on the life process of photosynthesis, food production, and greenhouses. The third would have to be the scantily-clad deeply-tanned young woman who asked me if I liked what solar energy had done for her. (No comment.)

A SOLAR UTILIZING SYSTEM

A basic solar usage system functions by collecting, transferring, storing, and using energy. Additional elements for shading, venting, and otherwise controlling the movement of this energy may also be required. Commonly, a solar utilizing system is composed of parts which perform these functions and are named accordingly—the collector, the heat-transfer medium, the storage system, etc. (see Fig. 2-1). In some systems, these parts are distinct and in others, like the Trombe wall, they are not. However combined, each functional part has its own responsibilities which should not, in any blend of aesthetic or practical considerations, be compromised.

The collector does what its name implies—collects solar energy—and a lot more. It must be oriented and angled to intercept the maximum amount of the sun's rays during the day; this is a subtle task and is the subject of several sections coming up. Being out in the sun also means being out in the weather; the collector must survive the effects of rain, extreme temperatures, and wind. In some climates, there are little extras like hail, snow, errant baseballs, sandstorms, jungle rot, earthquakes, tornados, and stampeding rabbits to contend with. These variable conditions determine the options available to the builder/

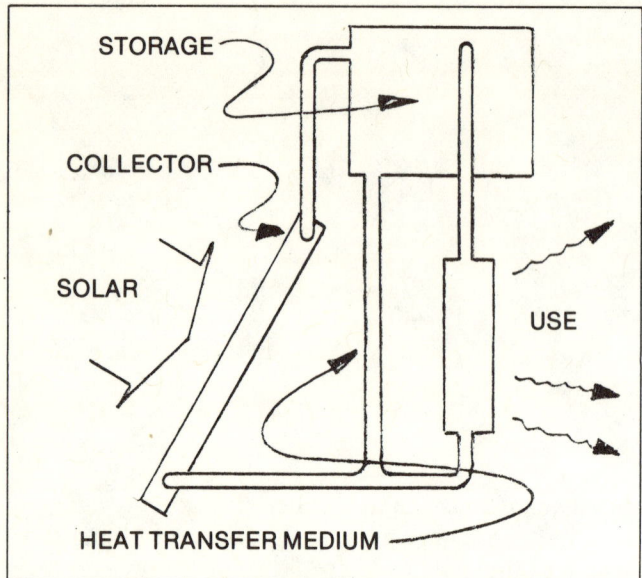

Fig. 2-1. A basic solar utilization system

buyer in materials, design, or the type of collector used for any given application.

Once gathered, the energy must be held; heat is very leaky stuff. Since it's only a question of time before the collector loses this battle, it must transfer the heat to useful work or storage when it is first, or best, able to do so. It's the responsibility of the heat-transfer medium to do this. Whether air, water, or glycol (antifreeze), the medium must absorb the collected heat and move it to immediate use or to storage. If correctly designed, storage is a maximum-security prison for heat, holding it for future use and allowing very little of it to disappear.

That's it in a nutshell. Making it all happen, of course, is something else altogether. The direct use of the sun's energy usually means conversion of the solar light into heat, but that heat is slippery stuff, about as likely to stay around as a plateful of cookies in a roomful of children. But heat isn't sly and it doesn't think; we can stop it from going where it might want to go. First, though, we have to know where it wants to go, and what means it may choose to get there.

HEAT PATHS

Our use of solar energy often requires that we handle heat. One very basic fact that should never be confused is: *Heat does not rise*. Thinking that it does or can will get you into trouble, and hinder a clear understanding of heat-transfer systems. What does heat do? *Heat seeks cold*, period! It's amazing to me that there are no heat cults anywhere in the world, because there is nothing that seeks oneness with the cosmos like heat. The physics term for this is "entropy," and it means the tendency of less randomness to travel to more randomness, for higher to seek lower levels, and for uniqueness to achieve uniformity. It's all part of a winding-down process that's been going on for a long time and will continue, let's hope, for some time.

While heat itself does not rise, in our gravitational system, some heated things *do* rise: air, for instance, as in hot-air balloons; water for another—one of the reasons houses and water tanks heat from the top downward. Sure, the heat source may be at the bottom of the tank, but as soon as the Btu's excite the water, the heated water goes to the top and relatively cooler water flows downward to be heated in turn.

One other hard fact of which you should be aware: Cold is the *absence* of heat. Too often, coldness is thought of as an entity in and of itself, so that it's a slug-it-out battle when you open the door of a warm house in winter and allow heat to fight the cold. Not true! Cold air, being a heavyweight, flows inward while lightweight hot air slides up and out. Because of the tendency for cold air to fall and heated air to rise, we can operate things like open-top freezer cases in the market, because the cold and hot air will stratify, forming distinct layers just as a water-and-oil mixture will. Heat is heat, and cold is the absence of that heat, and temperature is a measure of heat. When we remove the very last ounce of heat from something, it achieves absolute zero or about minus 460°F. Cold and hot take on new meanings in this light—it's all relative. Cold to a Minnesotan and a Hawaiian means different things. Referring to heat by degrees (this room needs 15 more degrees) may be helpful at times, but Btu's (British thermal units) or metric calories are terms we must acquaint ourselves with if we are to be comfortable working with solar energy. More on this later. . . .

Before we get into the thick of things, it's helpful to know that heat has only three forms of transportation: radiation, convection, and conduction.

Radiation is space-age stuff. It's the way that the sun gets its energy to us through a lot of vacuum. When you're sitting in front of a warm fire, what you feel is radiated heat. The air may be cold around you, but you're bathed in warmth. Radiated heat keeps going until it's absorbed, reflected, or transmitted through something; we'll be coming back to this point in a few minutes.

Conduction is another way that heat transfers itself, this time through objects. If it is easily able to do this, the material it travels through is referred to as a "conductor." If it has a hard time, we've got an "insulator." Knowing these two things, if we want to control the whereabouts of heat, we freely allow it to conduct itself wherever it is we want it to go or we greatly impede this flight by using insulation. The very best conductors and insulators are also very expensive, so we compromise. We still have many materials to choose from, some natural and some manmade.

Convection is the third way heat can move. This gets back to heated air or heated water, which, in the earth's gravitational field, rises if allowed to do so. The trick here is for the heated substance to stay in its own medium. No matter how hot you get that rock, it's not going to rise up in the air. But heat water, and it will rise through the surrounding cooler water. There's no magic involved. A heated substance expands, becomes less dense, and rises. It's from this effect that people get the notion that heat rises when what they really mean is that hot air or hot water rises.

Take the air or water up into orbit around the earth and neither will rise. There's an old astronaut's trick of striking a match inside a space station and, as it's held still, it dies. Then the guy who lit it starts yelling that there's something wrong with the air, creating a small panic among the newly arrived raw recruits. In reality, the match is extinguished by its own exhaust gases, since they cannot rise away in zero gravity!

So when heat gets confined against its will, it can only break away through radiation, conduction, or convection (see Fig. 2-2); account for all three, and you can contain it. Once liberated through radiation, though, this energy has a number of things that it can do. It can keep going, or, when it strikes anything, it can be transmitted, reflected, or absorbed (see Fig. 2-3).

If solar light hits a material and passes through, we say that it has been transmitted. Air passes the sun's rays, as does a vacuum and, to some extent, water; all of these transmit the light. A more precise definition indicates passage through a boundary of two media such as air to water, or air to glass. In solar lingo, the word "transmittance" is used to define how good a material is at passing solar light. For example, Mylar has a transmittance of 85%; this implies that 15% is absorbed or reflected. In actuality, anything which allows the passage of something else through it can be said to have a value of transmittance. For our purposes, that "something else" is the sun's light.

If radiated energy strikes a surface and bounces off, we say the energy has been reflected. This may not be a property of the material itself, but of the color of the material or the paint applied to its surface. White reflects much more energy than black. A mirror will reflect even more than a white surface. The angle at which the radiated energy hits a surface is also important for figuring how much is reflected from an object. If the light strikes the surface normally (perpendicular to it, or, as the kids would say, "head on"), less light will be reflected than if it hits the surface at a 45° angle. If light strikes even a blackened surface at a low angle of incidence, say 10° from parallel (a glancing blow), almost all of the energy will be reflected. A smooth surface reflects more light than a rough surface of the same color. Any energy that is reflected still has speed, but a new direction; this will keep going until further contact is made.

When the energy isn't reflected or transmitted, we say the energy is absorbed. In a way of speaking, it "melts" into the material that has intercepted the energy. This material might be warm or hot to the touch, because the energy has been converted into what we know as "thermal heat." If there is a cooler

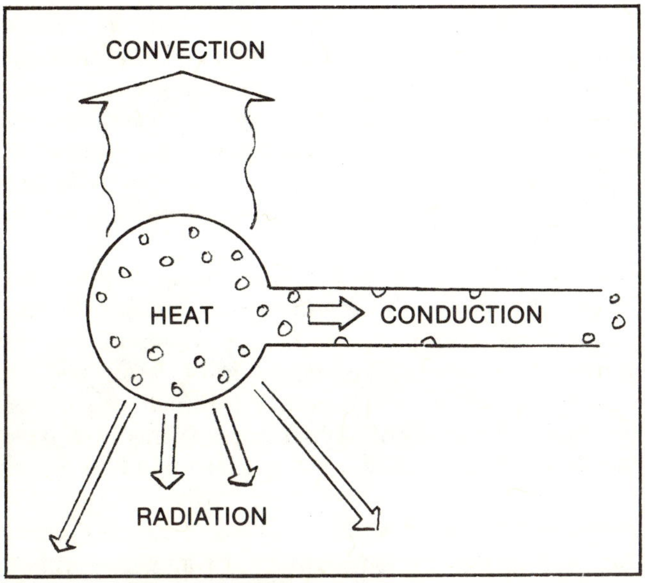

Fig. 2-2. Three ways heat moves

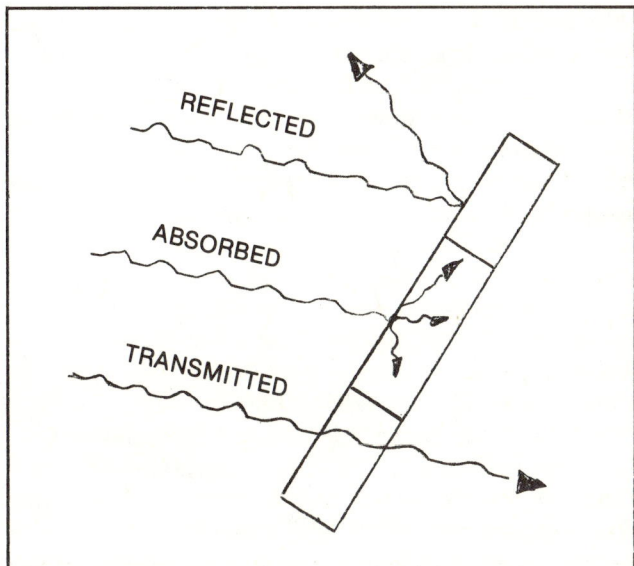

Fig. 2-3. Radiant energy impacting an object

place to go, the energy will immediately begin conducting or convecting its way there; or it may itself radiate. Only this radiation isn't the same stuff that came from the sun. If we placed a vacuum nearby, we'd see that this stuff could *not* pass through it. Nor will it pass through the glass that passed the sun's rays only moments before. Very strange. What is it? Well, it is heat, but it's longwave or low-frequency energy. That's just a fancy way of saying that it doesn't vibrate as much as the original solar energy, which is high-frequency, shortwave energy. Stand in front of a window on a sunny day, and you will easily feel the heat. But put a pane of glass between your face and the fireplace, and you will no longer feel the heat. Make a note of this unique property of glass: it's transparent to the sun's rays, but opaque to thermal heat.

Shortwave and longwave. High and low frequencies. What do these terms mean? Our solar light is much more than anyone, at first, imagined. What we see of this light—the visible band—is just a small portion of the wide spectrum that impinges on us and our environment. The sun, being very hot, emits a wide range of light frequencies, which indicate the varied vibrational states it undergoes. Only the higher frequencies—the shortwave stuff—can travel through a vacuum. Once it hits our atmosphere, most of the ultraviolet portion of this light is stripped away. If it weren't, life as we know it would be quite different. That which makes it through our atmosphere is what will strike our solar collector, when we've got it built. And certain frequencies are more useful for producing thermal heat than others. This range, for your book of trivia, is 0.1 to 2.0 microns and 3.5 to 5.0 microns; a micron is one-millionth of a meter and is a measure of the length of one wave of an incoming frequency.

What we should understand is that materials which exhibit transmittance, or reflectance, or absorption can do so only at specific frequencies. Glass, as we've noticed, has an affinity for transmitting shortwave energy (coming from the sun) and visible light (we can see the fire in the fireplace through it), but it balks at passing longwave energy—the heat from the fireplace, for example. In the visible band, an object may absorb most frequencies, but it's the one that it does *not* absorb which is reflected to our eyes and perceived as the color of that object. Now, there's an oddity! We call an object blue when that's the only frequency of light it rejects! White reflects all of the frequencies and black reflects none of them.

To use the energy of solar light we must somehow trap or gather it. This rules out the use of reflectance or transmittance; absorption is the only one that changes the solar light into heat. Since black seems to absorb the most, the absorber portion of our collector will be blackened, right? It's not the most perfect absorber—it must be reflecting *some* light if we can see it—but it will do. This is a start, but there's very much more that we need to know about collectors; let's get to it.

PASSIVE AND ACTIVE SYSTEMS

When mingling with the solar gang, you'll hear the words "passive" and "active" mentioned a lot. Pin any member of this club to the wall and ask for a definition, and you may get a variety of answers. Some seem to think that a system with distinct parts —collector, storage, etc.—constitutes an active system, and the tasteful merging of these components into the architecture of a house defines a passive one. Others disagree. Their claim revolves around the use of man-made or natural heat transfer methods; if pumps or electrical controls are present, that's active, and the systems without such controls are passive. Still other people define it in terms of whether or not the system requires a person's presence; if it's automated, it's active, and, if not, it's passive.

Along with these classifications, there are the exceptions and borderline systems, and the systems that don't fit the definition of either. I belong to a seemingly small group that is getting rather tired of the numerous connotations both names have. If I'm asked (and I'm usually not), I'd say that active and passive systems are defined according to the involvement of outside power. Any system using any kind of control, electric pump, motor, or blower which consumes energy is an active system. If the power for these devices is made on-site, however, it belongs in the passive category with any other kind of system which relies upon natural convection or heat transfer. This would, I'm sure, not be acceptable to the bulk of the people that use these words, but that doesn't matter. What does matter, however, is *your* own definition of what they mean. Just be certain that whomever you're talking with is using the same definition. If you suspect not, define yours and ask for his or hers. To disagree in belief is tough enough to handle; to disagree from a difference of definition is a waste of everyone's time.

HEAT CAPACITY

If we want heat to work for us in any application, we must first assess the quantity and quality of heat dictated by the situation. For example, in heating water for showers, dishes, laundry, etc., we are always dealing with a quantity (number of gallons) and quality (degrees of temperature). A medium-sized family might normally use 70 gallons of 140°F water per day. Cooking with solar heat, however, is entirely different. We must deal with temperatures above 212°F, because this is required to bring about certain chemical changes and to vaporize the water in the food; anything else is just warming the food.

A third example is space heating. Very few people enjoy being in a room that's hotter than 80°F, even in the winter. This is a very low temperature, easily achieved by even the simplest collector. However, what's crucial to space heating is not the quality of the heat (below 80°F), but the *quantity* of heated air. In the choice of collectors (soon to be discussed), one type works better when we're dealing strictly with the quality of heat the application demands, while the other type is better suited to a lower quality but larger quantity. And, as you will discover in the following sections, we should never get hurried in conversions; that is, use the "lowest-quality" collector that will do the job. Don't use a collector designed for high temperature in a low-temperature, high-quantity job!

COLLECTOR TYPES

It should seem obvious that, with so many different applications—space and water heating, cooking, dehydration, distillation, refrigeration, etc.—we need more than one type of collector. A collector for cooking food is very different from one needed to heat water. What may come as a surprise, though, is that there are only *two* basic types of collectors—the absorber and the concentrator—and only a few variations of each for the multitude of tasks they can, and do, perform!

ABSORBER COLLECTORS

Frequently referred to as the "flat-plate" collector, the absorber collector operates by the "body-on-the-beach" principle. Just lay it out there, and it soaks up rays like everything else. Being a collector, though, it's more efficient at absorbing the sun's rays than other items in the immediate environment. How? If you expose your pale body to the mighty sun god out on the beach sometime, you might think that you're a darn good collector. Stagger back to your car, which you left with all the windows up to keep from getting ripped off, and try to sit on that black upholstery and breathe the 180°F air, and you'll see how much more efficient it is than you are at collecting the sun's rays! That goes for the melted sunglasses on the dash, too.

If we all experienced temperatures throughout the year in the vicinity of 110–120°F, there'd be little need for a special unit to capture the sun's energy. We'd scramble the eggs on the sidewalk and take a shower from a hose that had been lying out in the sun. A collector, then, is designed as a full-time device, gathering the sun's energy when it's available, irrespective of the ambient (outside) temperature. It quickly achieves temperatures beyond those experienced in the environment, mainly because of its designed purpose—to hold the heat.

In the absorber collector, we might be called upon to heat the object which needs the heat (such as food), or we may heat water, air, or some other substance that, in turn, will heat something else. In most cases, even though we may need to heat air or water which is moving through the collector, we may first be heating some collector material (like steel, aluminum, or copper) that makes up the absorber surface.

There are some necessary parts to the collector besides the absorber surface—the container, glazing, insulation, a heat-transfer medium, etc.—but we'll get to those in due time.

Water-medium types. Let's look at some different styles of absorbers which use water as a medium —the twin-sheet, tube-and-sheet, drip, and pool. You may hear them by another name from another source, but I believe, along with the offshoots and combinations, that these represent the basic styles.

Twin-sheet. The twin-sheet absorber does for water what the single-sheet collector does for air. The two sheets, one or both usually corrugated, are fastened together and filled with water. Blackening one side of the sheet and putting this face up in a box covered with a glazing (like glass), we have what might be called, after several hours in the sun, a heated water sandwich (Fig. 2-4). Inlet and outlet connections are made at the bottom and top (respectively) for moving the water to storage.

The twin-sheet absorber doesn't see much application in water-heating collectors anymore. Whether one or both of the sheets is corrugated, it's difficult to seal the ends to hold in the water. Two flat sheets, however, present a problem in maintaining a distance

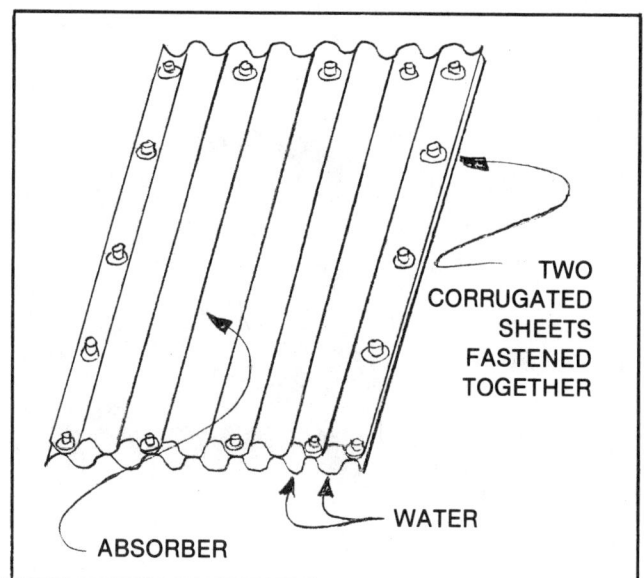

Fig. 2-4. Twin sheet absorber

between the sheets for the water itself. Attempts to use spacers or drill holes to secure the plates together rapidly create sealing problems, and the whole affair ends in a designer's nightmare. Solution? Avoid it. I only mention it so you'll think twice if someone tries to pawn his mistake off on you at a bargain price!

Tube-on-sheet. This type of absorber collector is very popular (see Figs. 2–39 through 2–43). Two new innovations—the tube-in-sheet and the tubed sheet —also provide the same service, but since they require fairly sophisticated techniques, jigs, and machinery, they will not be discussed. The principle behind all three, however, is simple: the sheet's material absorbs the sun's energy, and the tube portion contains the heat-transfer medium, usually a fluid—water, glycol (antifreeze), etc. A later section of this chapter looks at some of the basic tube-on-sheet arrangements. In each case, however, it's important that the heat, once absorbed, flows through the sheet and through the tube wall into the heat-transfer medium. Since there are many places the heat wants to go, we must ensure the least amount of resistance to its taking this particular path. Therefore, the sheet and tube materials should be excellent heat conductors. Copper sheet and copper tubes, however, aren't enough; the heat must get from one to the other. This is done by making a good connection between the two, and it's one of the reasons that the tube-in-sheet and tubed-sheet were developed; one gives a perfect mechanical connection and the other "tubes" the sheet itself so the two are really one. For the tube-on-sheet, a good connection is practically impossible, since so little of the surfaces of each come in contact. To improve the connection, the tube is "bonded" to the sheet.

For copper, a popular bonding technique is soldering, which is terrific if you can afford copper sheet. More frequently, copper tubes are used with aluminum or tin sheet, or aluminum tubes with aluminum. For these combinations, solder won't work—besides, it's a no-no to mix metals, since their dissimilarity can lead to corrosive (galvanic) reactions. Getting a bonding agent to be thermally conductive but not electrically conductive wasn't easy until thermal bonding cements came along. Attempts to use resin-type agents uncovered a further complication—the bonding agent had to withstand thermal shock (a sudden inrush of cold water into the hot collector) and differential expansion rates between the dissimilar materials. See the Sources and References section for a few cements that fit the bill.

Too many tubes on the sheet means a high-cost absorber, but too few tubes means large expanses between the tubes; heat absorbed there would just as soon radiate or conduct its heat to the air in the collector as travel all that distance to the tubes. A happy medium seems to be 6–8 inches; if you're using corrugated aluminum or tin, that comes to every-other-valley.

The level of your skills and whether the system is pumped or thermosiphoned (we're getting to this soon, so read on) will help you select the best configuration of the tube-on-sheet for your system.

Drip. An interesting technique for gathering energy from the absorber-type collector is to allow water to run down over the heated absorber and heated in the process (Fig. 2–6). The water is first pumped to the top of the collector, where it exits from small holes drilled in metal or plastic pipe. Gravity then takes over and the heated water is chan-

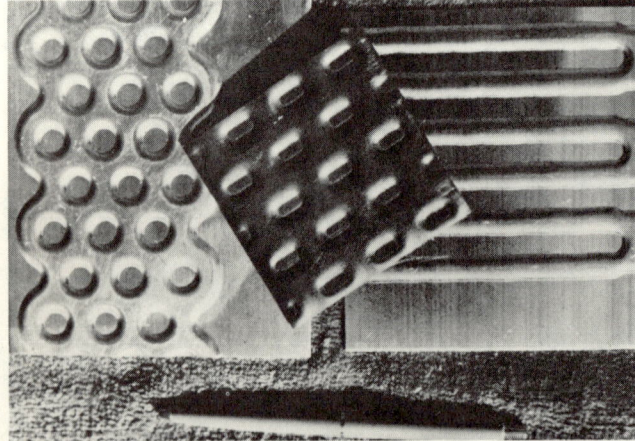

Fig. 2–5. The absorber surface on some commercially available solar collectors is "tubed-sheet."
(*Courtesy James L. Ruhle & Assoc.*)

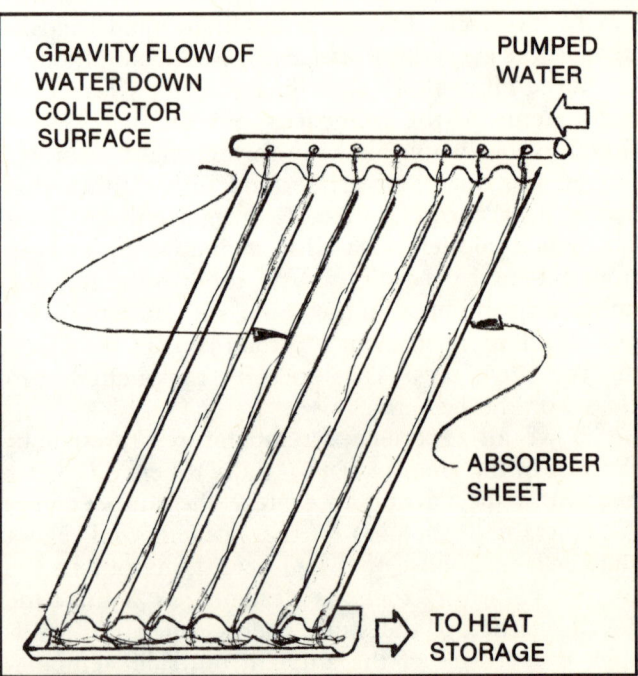

Fig. 2–6. A drip-water absorber

neled into the heat-storage area. This concept is a good one for very cold climates where any water in the collector might freeze several times a year. In the tube-and-sheet arrangement (if water is used), the collector must be emptied when freezing conditions are reached—an irksome task. An automatic sensing system for this can be expensive, and how do you dump water that's under pressure? In the drip system, the sensors simply turn off the water pump and there's no more water in the collector. Simplicity! Exposed water in the collector presents its own set of problems, however, limiting the types of material that can be used for the glazing and collector container. The absorber surface has a propensity toward growing algae and other green things, too. Frogs plopping about in the collector won't do!

Pool. The pool absorber collector is easily demonstrated on a large scale. Next time your mother-in-law's swimming pool is drained, paint the bottom and sides with a good flat black paint, and refill. Great for helping heat the pool water, because the sun's rays travel through the water, get absorbed by the black surface, and immediately conduct their heat to the water. Without a few other provisions, however, this is not recommended. At nighttime, this pool would demonstrate another principle equally well—night-sky radiation (see the section later in this chapter on cooling techniques,—guaranteeing a nice ice rink in some pretty summery weather. Besides, swimming pools with black bottoms and sides are weird—and very scary to swim in!

Do this on a smaller scale and you'd have a good solar still. You'd think it would give you some heated water, but it doesn't. As soon as the water starts to heat, it begins to evaporate, rise, and condense on the glazing, defeating the purpose of the collector! If you eliminate the air space, you bring the water in contact with the glazing, and it just as readily gives up its heat. The pool absorber, then, is strictly for distillation. Slope the glazing material so the condensed fluid runs down into a tray (separate from the pooled water) and draw off that water—or whatever it was you were distilling.

Air-medium types. An air-type absorber collector is one which used air as a heat-transfer medium rather than water. As we can see from a glance at Fig. 2–7, the outward appearance of the collector is identical with the water-medium units. Inside, however, nothing could be simpler. A blackened absorber still absorbs the sun's rays but, with no need to conduct the heat to tubes of water. The energy is used simply to heat air which comes in contact with the absorber surface. This means, among other things, that the absorber material need not be a conductor!

This represents the single-sheet, or so-called "flat-plate," collector, casually used to refer to the general class of absorbers. This is misleading, since other styles of absorbers exist which look nothing like a flat plate. Watch that nomenclature.

While a flat sheet of metal, plastic, or wood may fit the name of the collector, a corrugated sheet may, because of its stiffness, require less support in the container. Wood is a good absorber material, but corrugated sheets of aluminum, tin, or galvanized steel are also useful.

There are a few neat arrangements for the air-type collector that may help mate this unit to your system, and a few that will trouble the unwary. Look 'em over; we've got the single-action, double-action, and window-type collectors, finishing with the beer-can special.

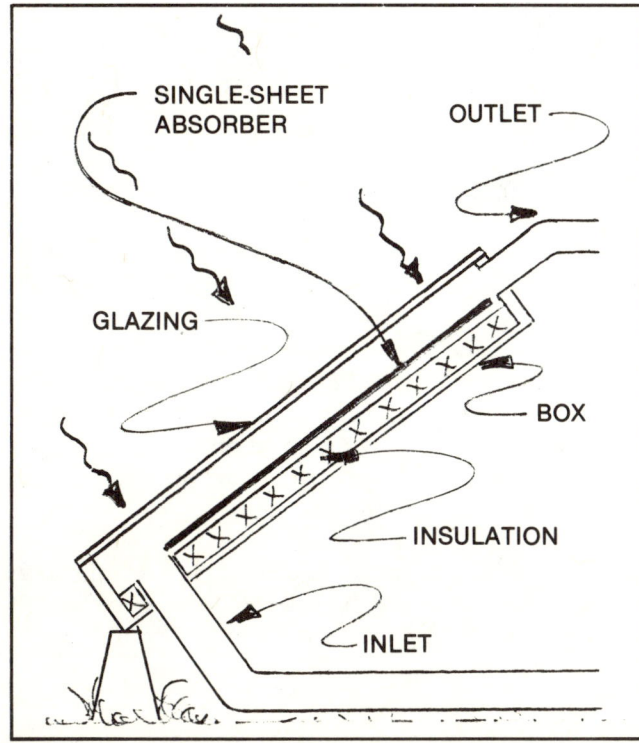

Fig. 2–7. A single-sheet, air-medium absorber

Fig. 2–8. Solar-heated houses can be aesthetically pleasing

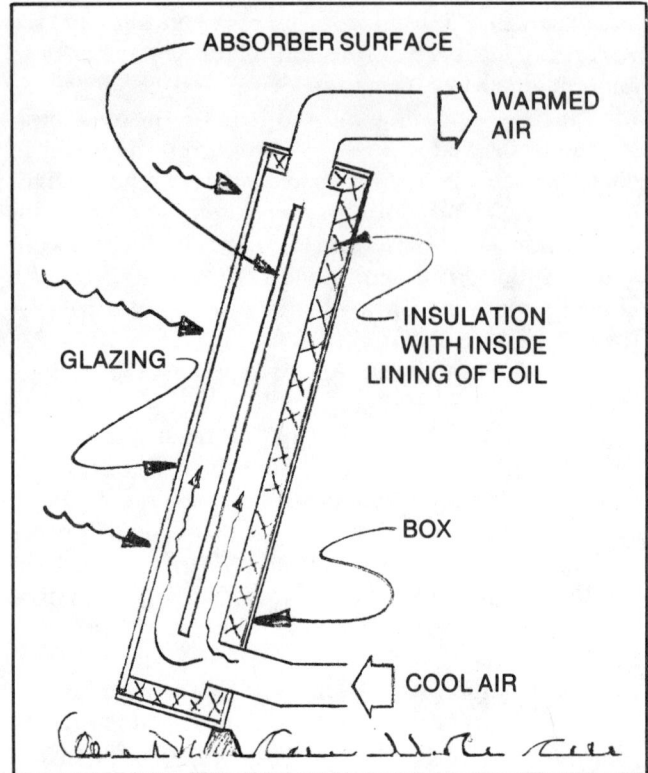

Fig. 2-9. A "double-action", air-medium absorber

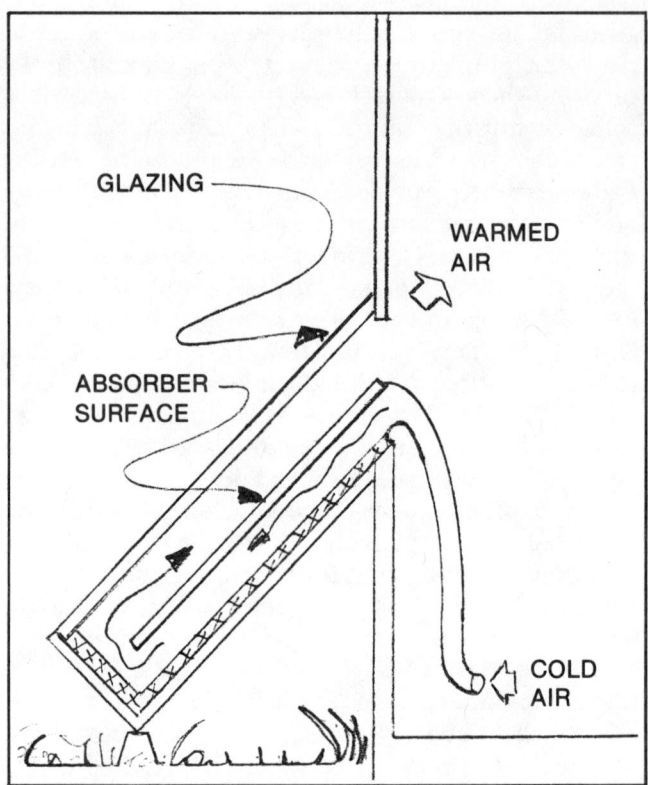

Fig. 2-10. The "window", air-medium absorber

Single-action. The single-action collector is a single-sheet unit. The warmed air, thermosiphoned or blown from the collector through the discharge air duct, sucks in cooler air at the intake located near the bottom end. The intake duct is usually attached behind the collector near the bottom. This avoids a tight bend of the ducting when trying to go through the bottom plate. Also, since the collector may be rested on its bottom (instead of "hung"), it avoids the obvious interference a duct would cause. Finally, collector design is liberated from the air-duct diameter as a factor influencing the collector box's depth.

Double-action. The double-action air collector (Fig. 2-9) recognizes that conduction losses through the backside of the absorber surface need not occur. Therefore, the absorber is suspended or mounted in the box with an air space behind as well as the one in front. Likewise, the incoming air is encouraged to seek either route, traveling up the front or the back of the absorber. The only other necessary modification is to put a reflective substance on top of the box's back insulation to repel radiated heat from the absorber's backside.

Window. The window air-type collector seems to have evolved from the need for a quick warmup of bedroom, bathroom, or kitchen air temperature as the sun is breaking over yonder mountains, trees, or high-rises. Avoiding the intensive system, some-

thing was required that would fit into a window, preferably without taking away the nice view. Enter the window-style space heater (Fig. 2-10). Sporting a low profile, this is attached at the window sill and braced out from the building at the bottom. With inlet and outlet at the top, this design embraces the best attributes of the double-action collector, except that the movement of air on the backside of the aborber material is downward. Any heat which has slugged its way through the absorber's backside, therefore, preheats the intake air for the sun's side of the absorber sunface.

This is a hard-working thermosiphon unit. If you want to pick up the cold air from the house floor, add an extension to the intake duct. If you use stovepipe-type (round) ducting, some of the snaky vent tubing used with clothes dryers will fit and work nicely. Prepare yourself for the comments this setup is likely to draw from visitors. After all, it does look as if your solar collector has its snout down on the kitchen floor, snarfing around for food scraps!

Beer-can special. There's a nasty habit connected with air-medium collectors, particularly thermosiphon systems, which I call the "whoosh-whoosh" effect. This is the point when the heated air finally discovers it has a place to go, and goes there—usually with a whoosh. In collectors with a long narrow channel (like our shop's roof system), there's an initial whoosh, followed by a waterfall of air that

SOLAR ENERGY

Fig. 2–11. An air-heating collector made with beer cans

sounds suspiciously like a tornado. Unchecked, this column of air can attain some pretty good velocities in startup and, under the right conditions, in everyday use.

Overcoming the propensity for air to move rapidly through the collector is a must for efficient heat transfer and noiseless operation. This is accomplished by implanting obstacles in its path, resulting in turbulence of the warming air. A simple turbulator is a beer can, cut in half or quartered, which is glued, stapled, riveted, or screwed to the absorber surface. If you're using metal for the absorber, a bonding cement might prove effective. Position the quartered cans in a random or diagonal pattern (Fig. 2–11) to resist laminar flow—the more turbulence, the greater the air-absorber contact and heat transfer.

Cutting and attaching all those beer cans might sound like a real job but, from personal experience, I can verify that the time goes quickly. After all, the cans must be emptied beforehand, no? It is never difficult to find help in this chore. Wait a day before doing any work roofside, however. One mustn't worry the family!

Air vs. water. Air-medium collectors are simple, straightforward, and clean. No water to leak or freeze. No expensive glycol to add or to worry about getting in your drinking water. Minimal weight in the collector. Air is very light stuff. Less thermal mass and more materials to choose from for the absorber surface. No soldering, tube bending, drilling, sheet-metal cutting, tube hacksawing, or tube jigs. So . . . what's the catch?

Intrinsically, simple collectors do simple things. You'd want to put away that window-size air-medium collector when summer arrives; you'd no longer appreciate its early-morning effect. Larger air-medium systems can be very competitive with their water-medium counterparts. Most air-medium units, however, cannot heat water to the degree any of us would consider useful for showers, washing dishes, etc. The air collector's main function, then, seems to be reserved for space heating. The kind of storage that air-medium systems require further reduces their application in finished homes that won't accept the necessary modifications. More on this soon.

CONCENTRATOR COLLECTORS

Compared with the absorber, the concentrator collector is dynamic and energetic. As children, some of us first came to know the principle behind its operation when we discovered the Death Ray. You know—you borrow Grandmother's magnifying glass and, on a bright and sunny day, go searching for an unsuspecting ant colony. A good exercise for hand-and-eye coordination as you chase the scurrying critters over hill and sidewalk, leaving a searing trail strewn with crispy-fried ants. The fire-breathing beauty which remained cool to the touch is the parabolic lens, used in telescopes, binoculars, cameras, etc., and it functions by taking the rays which strike it over the whole of its surface and focusing them into a concentrated beam. Thus, the concentrator.

In the concentrator group, the parabola curve plays the lead role, sometimes disguised as the glass lens or the pancaked Fresnel lens. Before we discuss the differences between these two forms, note a similarity: both concentrate the sun's energy on the backside. Reverse the roles, and you have the concentrator dish or tray; the concentrated energy now appears on the sun's side. Take a small enough segment from any portion of these four, and you have a nearly flat surface, giving us a fifth concentrator—the flat reflector—which is able to concentrate the sun's energy toward, or away from, the sun as well as points between.

Parabola. Ask a mathematician what a parabola is and you'll get "a slice of a cone." Ask a gunnery striker and you'll get "the arc a shell makes when fired off downrange." So, a parabola exists in theory

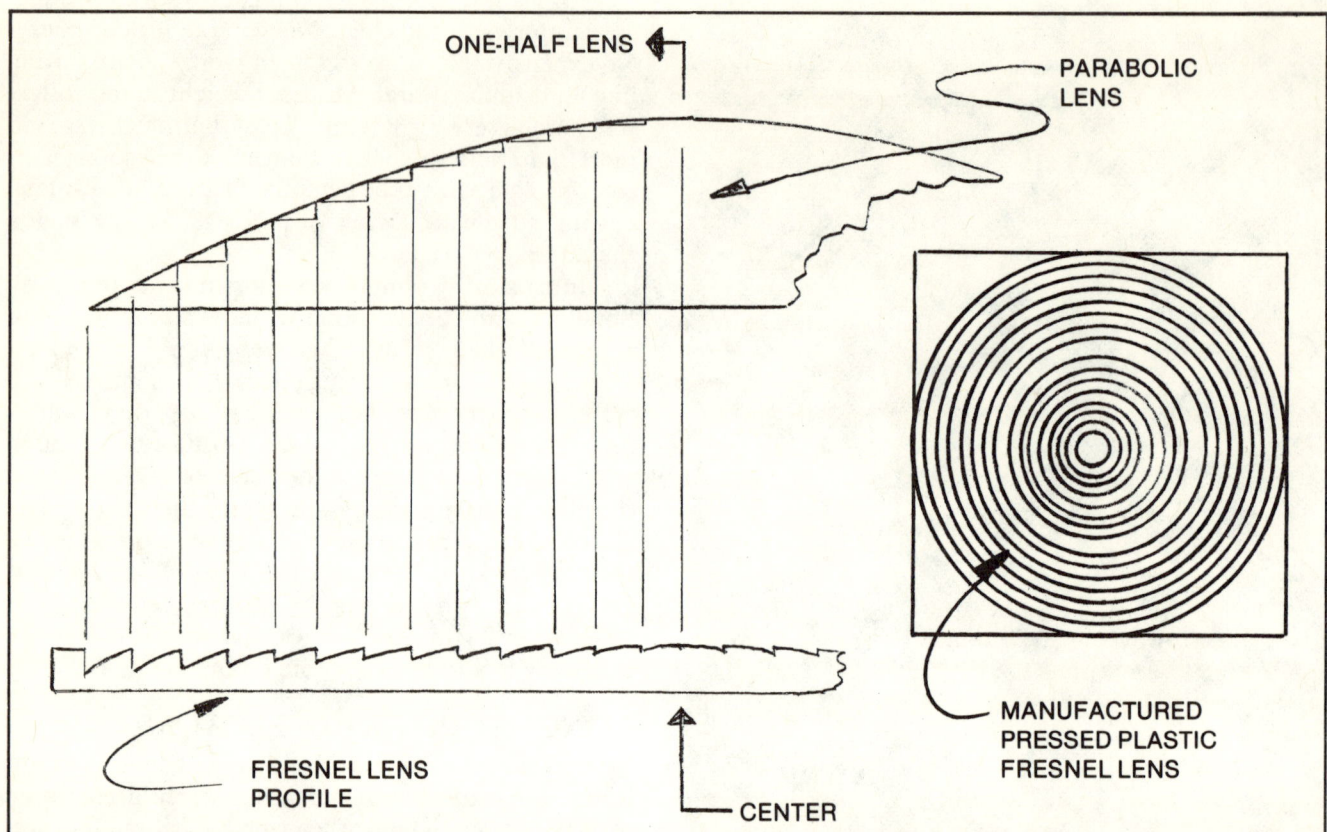

Fig. 2–12. The Fresnel lens

and in practice. But it shines its brightest under the sun. If you were told to generate a curve which would reflect the essentially parallel rays of the sun onto a single point, this would be it. Accordingly, if you put a light source at the focal point of a parabolic curve, all the light which is reflected from it would be parallel; that's how flashlights and car headlamps can "throw" light so far.

The absorber collector, while able to achieve some pretty high temperatures, can't compete with the concentrator; a mile-square absorber collector array, sitting in the hot sun for a whole day, cannot ignite the paper that a 2-inch-diameter parabolic (lens or dish) can when held for only a few seconds under the noonday sun. That's what happens when you cram quantity into a very tiny space; it suddenly becomes quality energy—the kind needed for some types of cooking, or generating steam, or melting metals. Power. Moxy. The old absorber is dull in comparison.

Parabolic lens. A lens that concentrates also magnifies. As if the world wasn't large enough, people want portions of it made a little larger. I guess you could say that concentrators got their start from poor eyesight. Reading glasses were parabolic lenses. Everybody knew how much trouble Galileo got into when he built his infernal telescope, so they found other uses for the things—capturing images on film, looking at germs and things. In the sun energy line, the lens saw limited action, since it quickly reached unmanageable size and tremendous weight. Somebody came along and used wine inside of glass "pillows" (and then, finally, plastic) to avoid the use of glass all through. Don't laugh; they worked! But there was the continual fear that a workman, knowing what constituted the innards of the lens, would tap it some dark night to pass the time. The parabolic lens (wine, water, glass, or plastic) was doomed.

Fresnel lens. A workable solution to the bulky lens is the Fresnel (pronounced fray-nell) lens. If I handed one to you, you would notice a few things. One, it is a piece of plastic about 1/8-inch thick and 12-inches square. Two, it has a lot of concentric cuts, or etches, in the surface on one side. And, three, if held in the sun, it performs exactly like a parabola! That's because it is a parabola—only it's a very special parabola. As you can see from Fig. 2–12, the thinness of the Fresnel lens is possible because we understand that the importance of the parabola is the angle of the curve, not the bulk of the material used to make it. By chopping the curve into smaller concentric segments, we get the desired effect minus the expendable mass. It's a marvelous innovation. Unfortunately, the only size Fresnel lens available is the one

I have described and that is commonly used for firing jewelry (glazing).

Dish. The parabolic lens and Fresnel lens refract the sun's rays to obtain a focused source of energy (light and heat). However, if we point the concave (inner) side of the parabola toward the sun, instead of the convex (outer) side, and line this surface with a reflective material, we get the focused energy on the sun side of the apparatus. This is the parabolic dish (Fig. 2-13). The reflective material can be glossy silver paint, chrome, a mess of tiny flat mirrors set into a parabolic bed, aluminized Mylar, etc. The reflective coating is often very thin, depending on another material of less cost, such as wood, concrete, or steel, to form the structure. Even though the dish concentrator produces a lot of heat, the components which make it up (unless in the close vicinity of the focal point) will not get hot.

Tray. The most widely used concentrator today is the parabolic tray, formed by a parabolic curve that is swept through a length (Fig. 2-15). Instead of concentrating all of the rays into a single point, as in the dish, the tray focuses them into a line which is as long as the tray itself, overcoming the main difficulty in applying the parabolic dish to any useful function. That is, high temperature (quality) is nice, but we need this heat in quantity, too. Increasing the size of the dish, however, only nets still higher temperatures. If you're trying to boil water to make steam, it doesn't help much to melt the steel pipes containing the water! Enter the parabolic tray; its width determines the degree of temperature of heat you'll get and its length gives you the quantity. But we're getting a little ahead of ourselves.

Flat reflector. Not all concentrators are parabolic. A simple flat sheet of a reflective material will serve adequately. Operating a signal mirror, for instance, demonstrates that, with a rather small device, we are able to put much more sunlight into a given area (the point at which it's aimed) than would normally fall there. That one should be able to see the flash of sun off a 2-inch-square mirror at 20 miles in daylight, and *not* see an automobile headlight (which uses a parabola) at a mile in the same daylight, indicates not only how parallel the sun's rays really are, but the effectiveness of the flat reflector.

The concentrating effect of the flat surfaces, however, is dependent on the number of flat surfaces used and/or their size. Flat surfaces, in and of themselves, do *not* concentrate the sun's energy. Rather, the sun's energy is "redirected" onto another surface; we've added the light that would have normally fallen elsewhere. With a number of reflectors also aimed at the same place, we have a concentrated area of solar energy.

Properly arranged, flat reflectors can put extra

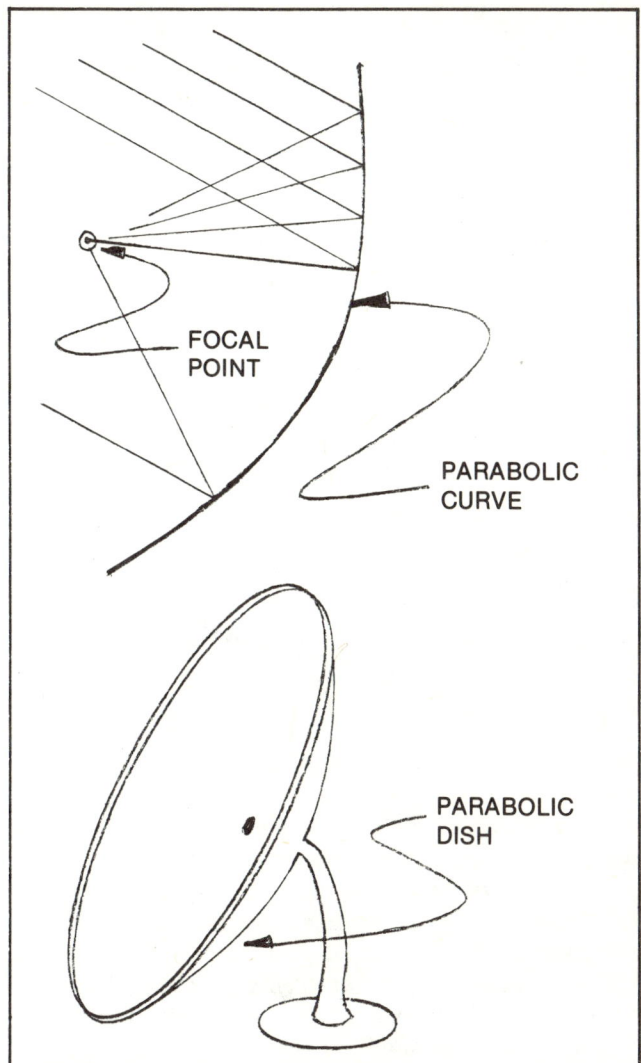

Fig. 2-13. A parabolic dish

Fig. 2-14. A solar oven with a multi-reflector design

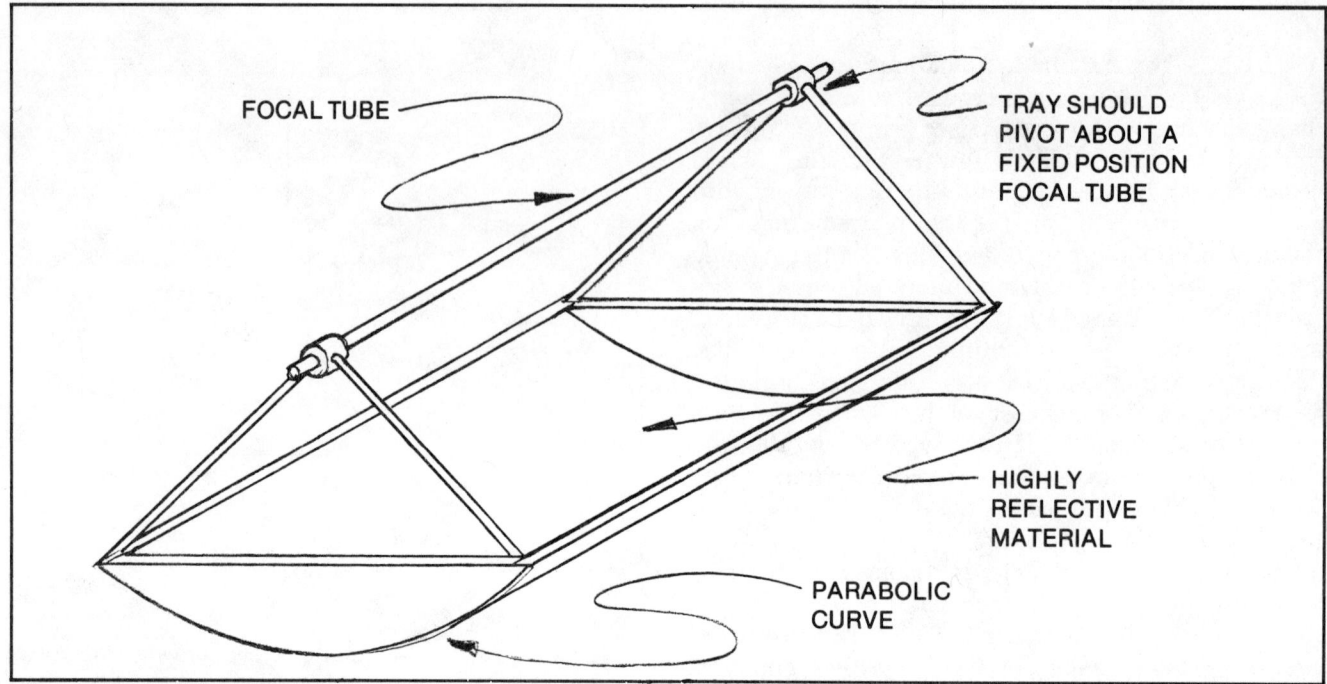

Fig. 2-15. A parabolic tray

energy into points in front, or back, of their own respective position (Fig. 2-16). Their best work is done when reflecting energy toward the sun side; the more perpendicular the sun's rays are when they strike the surface, the greater the projected area of reflectance. That is, while the surface reflects all of the light that strikes it, it gets less light because of its own angle. At 45°, for example, it gets only 70% of the light it would at 0°; this is reflected at a right angle to the incoming light. At a 60° tilt, only 50% of the light is deflected sideways and backward.

The flat reflector is sometimes used in conjunction with an absorber collector, but most often it's alone (see the discussion of solar ovens later in the chapter). The only time it's used with a parabolic collector is when the latter is too large to track the sun. In the heliostat arrangement (Fig. 2-17), the parabolic collector is faced to the north (in the northern hemisphere) where the flat reflectors are located. The reflectors face south and, with suitable tracking mechanisms, focus the sun's energy, throughout the day, into the stationary parabolic concentrator. This is useful on very large installations where it's essential to control the incoming energy. By varying the number of flat reflectors that stand idle and those that track, this setup can provide from zero to full sunlight to the parabola. Mega-heliomania.

FACTORS AFFECTING CONCENTRATOR EFFICIENCY

There are two major prerequisites to using parabolic concentrators. One is the presence of direct sunlight; if a cloud blocks the sun or if there is a light overcast, the concentrator collector will not work—it must have something to concentrate. An absorber collector, on the other hand, can work better on a lightly overcast day than a bright sunny one; it loves to take its energy from a sky full of reflected energy even more than from a single point source.

The second necessary ingredient to a parabolic concentrator's workability is sun-tracking; if it doesn't constantly point at the sun, the focal point shifts. Units which have a stationary concentrator and trackable focal points do exist, but by far the most common arrangement (and the least sophisticated) is for the tray to follow the sun throughout the day; it must do this accurately.

To simplify the process, a number of design factors are combined. We need to put something in the focal point of the tray if we want to take advantage of all of the energy concentrated there. Invariably, this is a pipe with a heat-transfer medium or working fluid. If we extend this focal tube through the ends of the tray and fix it in bearings or bushings, we now have a tube that rotates in a tray. Conversely, we have a tray that will rotate about the focal tube. If we orient this axis north-south and tilt it to the correct angle of declination for the time of year, we can rotate it until it points at the sun. In fact, we can "find" the sun at any time of the day. Since there are only about eight hours of worthwhile energy per day, the tray only needs to sweep through 120°. How did I arrive at that figure?

The earth makes a complete rotation on its own

SOLAR ENERGY 17

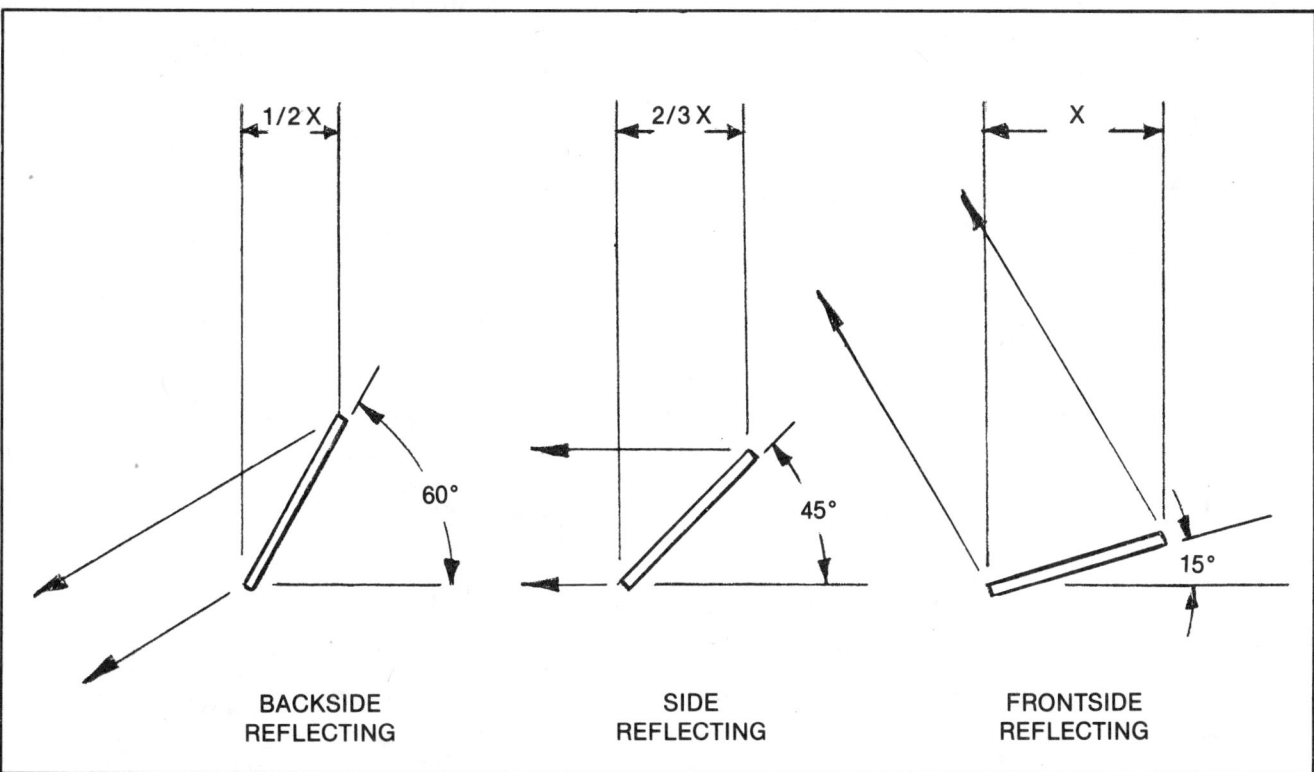

Fig. 2-16. The flat reflector

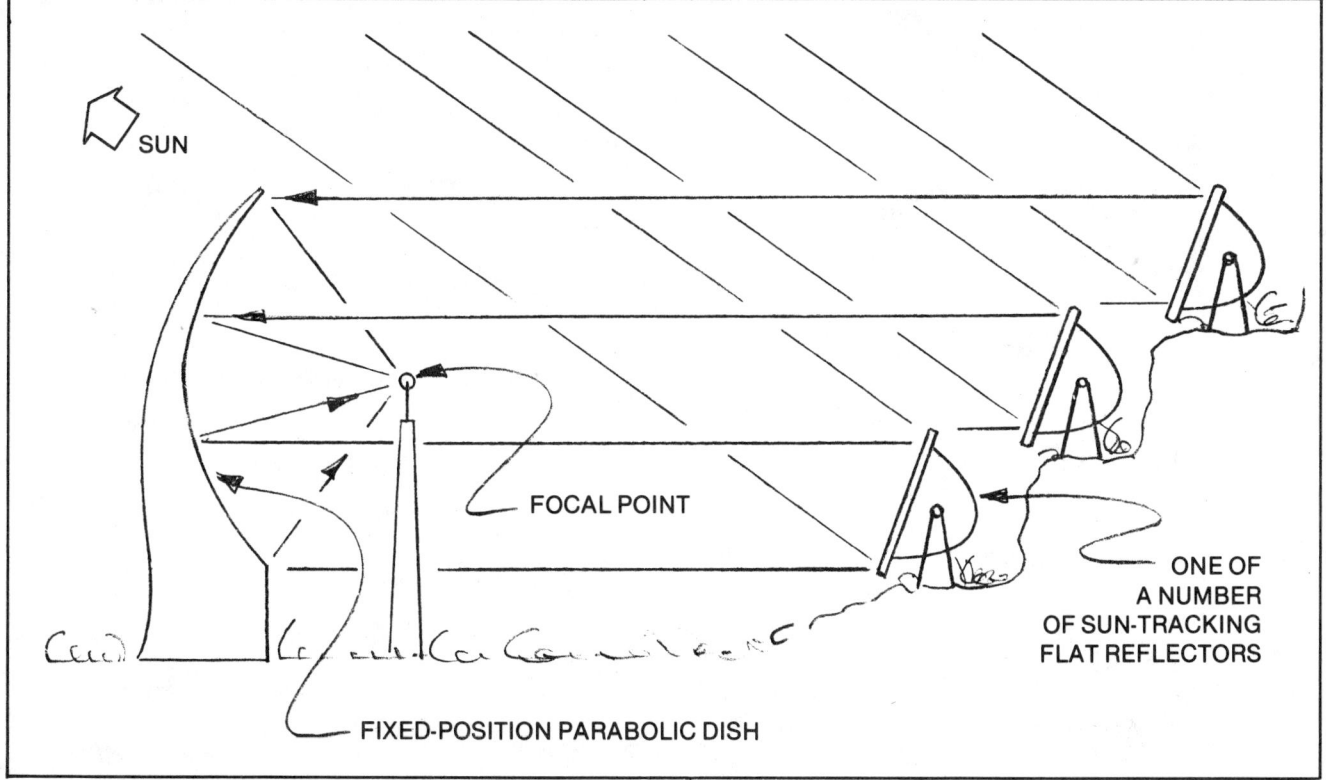

Fig. 2-17. The heliostat

axis in twenty-four hours, right? There are 360° in a circle, right? Divide 360° by twenty-four hours and you get the number of degrees the sun "moves" in one hour; it's 15°. If we have eight hours of tracking to do, that's eight times 15°, or 120°, that we must sweep with the tray. Since you'll need to know this if you're building a tracking mechanism, I'll ask the question: How long does it take the sun to move 1°? Here we divide sixty minutes (one hour) by 15°; we get four minutes. Or, in one minute, the sun "moves" a quarter of a degree. Fascinating, no?

One variable of parabolics is the focal distance, which is the measured space between the focal point and the nearest portion (center) of the curve. Don't go to sleep—this is important! We can vary the focal distance by varying the parabolic curve. And you thought there was only one parabolic curve! A flat parabola (broad curve) has a much larger focal distance than a sharply curved one. Whatever the application, we must select the focal distance (and, thereby, the parabolic curve) with care. Why? Three reasons. First, irrespective of the type of reflective surface (coating or material), few, if any, of these can stand even a minimal exposure to the elements. Therefore, like the absorber, the parabolic tray gets a glazing (transparent covering). Its job is *not* to contain heat (although that's a fringe benefit) so much as it is simply to keep rain, snow, dew, dust, and grasshoppers off the high-gloss surface. Focal distance now rears its head; we'll want a focal point "inside" the tray. Assuming that you've selected a width of tray (for aesthetics, best material use, or to achieve a specific temperature at the focal tube), you must now find a parabolic curve that fits in that space and yet allows closure of the tray (by the glazing) over reflective surface and focal tube alike. As indicated by Fig. 2-18, you've got a number of curves that do the job.

The second reason that focal distance is important now becomes apparent. A limiting factor on the number of curves that will fit a given width is tray balance. Considering that you must rotate the tray about its axis (which is the focal tube) to track the sun, it helps matters if you select a focal distance which gives you a proper distribution of weight. Side to side, this is done; it's just as far from the focal tube to one side of the tray as the other side. Up and down is a whole different story. Above the focal tube is the glazing and a portion of the tray (on each side), and below it is the rest of the parabolic curve and reflecting surface. So what you need is to pick a focal distance that balances out the weight, however distributed, above and below the focal tube. This is ultimately situational, so there's no rule of thumb. If your hypothetical parabolic tray uses glass for glazing, the focal tube is going to end up closer to the (heavy) glazing than the bottom of the curve. Once you've decided on the materials for the collector, figure this out.

The third factor affecting focal distance is the

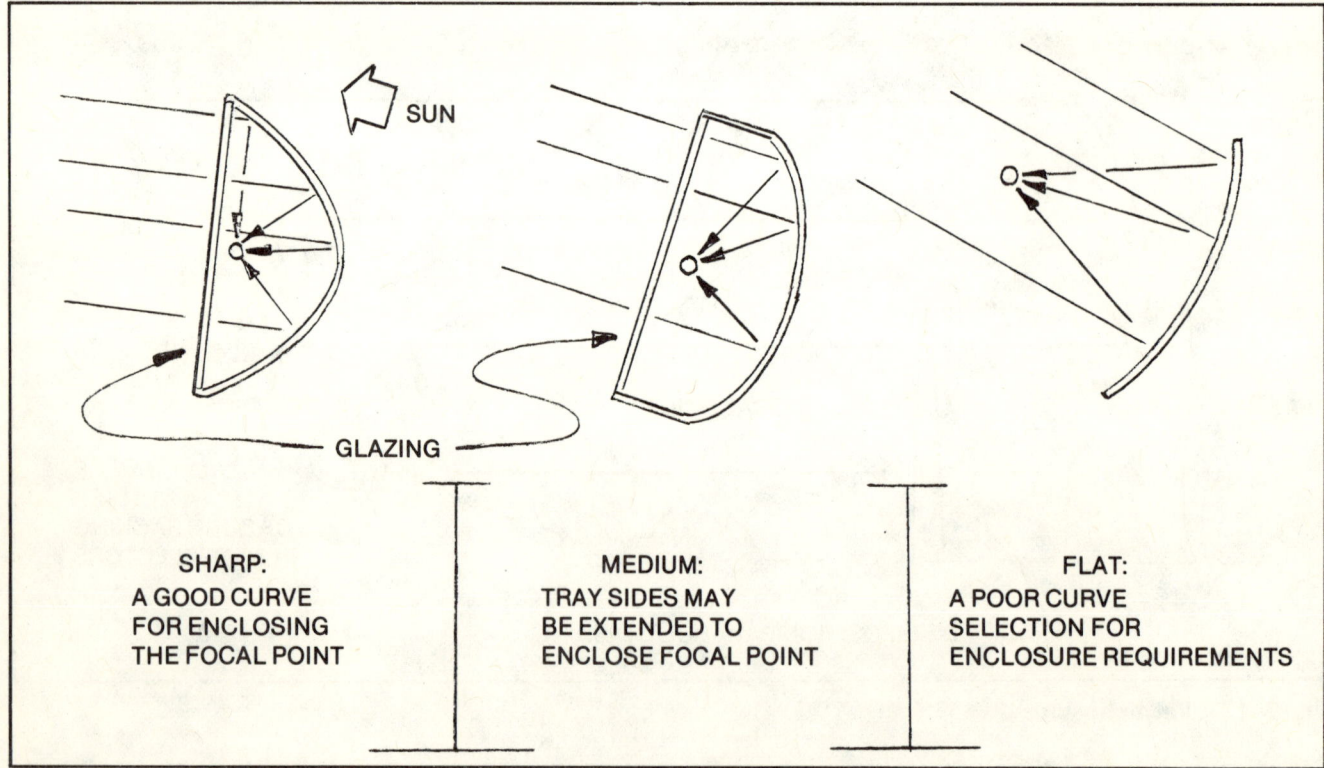

Fig. 2-18. Enclosing the focal tube in the parabolic tray

tracking mechanism itself. There are many designs, but the net effect is continuous drive or a pulsed advance. The continuous drive means a smooth, even, going-all-the-time tracking system. The pulsed advance is choppy, now-it's-on-now-it's-off stuff, and for the periods as long as four or five minutes, the tray doesn't move. The sun's position does though, and so does the focal point. Consequently, the focal tube will be in and out of the focal point. There are two ways to solve the problem. Use a larger-diameter focal tube; this is objectionable since a larger surface will increase the surface heat losses. Or, minimize the effect of a few minutes' swing of the sun by keeping the focal distance as small as possible. The larger the focal distance, the less time the shifting focal point will stay on it. If you can't immediately visualize this, take a baseball and, at arm's length in a darkened room, hold a flashlight beam on it. Note the amount of arc you can swing the flashlight and still illuminate the ball. Now, put the ball about ten feet away and notice the considerably shortened arc that still lights the ball. It's no different for the tray. You'll need to juggle tray balance, the tracking mechanism, and enclosure in your determination of focal distance and the specific parabolic curve generated from it. It's far from a chore, though; designing your own can be fun!

Sometimes we can't size the tray according to our rule of thumb—width supplies temperature, length delivers quantity. If you must use a width of tray wider than the maximum temperature you wish to attain, or if for any other reason you want to decrease the energy concentration at the focal tube, you can set it slightly out of focus. For any given focal distance, points on either side of it will be out of focus (Fig. 2–19). The energy is still there; it's just spread out a little more. If you use this method of temperature control, consider what is going to happen if your tray ever misaligns itself with the sun. A shifting focal point can quickly ignite materials it comes in contact with, so allow plenty of clearance in the tray for any degree of its sweep.

CAUTION: The focal point of a parabolic concentrator, no matter how small, is intensely hot—easily 1,000° or more, even with a 6-inch-diameter dish. Not exactly a nice place to put a finger or hand, right? You knew that? Well, did you know that the focal point of a parabola is more than just a point source of heat? That's right. It's concentrated light too. Stick a black object in there and it will not only get very hot, it will get very *bright*—blinding bright, as a matter of fact. When around concentrators, carry welding goggles. If looking into the tray or at the focal tube, *wear them*. As in welding, the effects are not always immediate; blisters and other signs of eye damage may not show up for many hours. Or worse yet, that tray may permanently blind you, which won't exactly make you a great advocate of solar energy. Last but not least, a focal point is *invisible*. We only see its effect when we put something into its focus, intentionally or not. "Eureka" is *not* one of the things you'll yell when you inadvertently discover its presence. It's silent, odorless, and (I'll assume) tasteless. Don't become a statistic!

MEDIUM TRANSFER

No matter what type of heat-transfer medium is used —air, water, glycol, oil, grapefruit juice, etc.—there are only two ways of moving it to storage or immediate use: thermosiphoning or pumping. That is, it moves by itself, or it gets pushed. Let's consider fluid transfer first and then we'll check out air-heat transfer.

THERMOSIPHONING WATER

In thermosiphoning, we are depending on the heated medium to "convect" itself away from the spot where it was heated, i.e., the collector. The reason it wants to do this is simple. When a fluid is heated, its molecules really start jostling each other, seeking elbow room, and they're pushed apart; we know this as expansion. If something expands, its original density decreases. Relative to the cooler medium which surrounds it, it's light. In a gravitational field, that means it will rise—in its own medium, of course; hot water won't rise through air. When it rises, some-

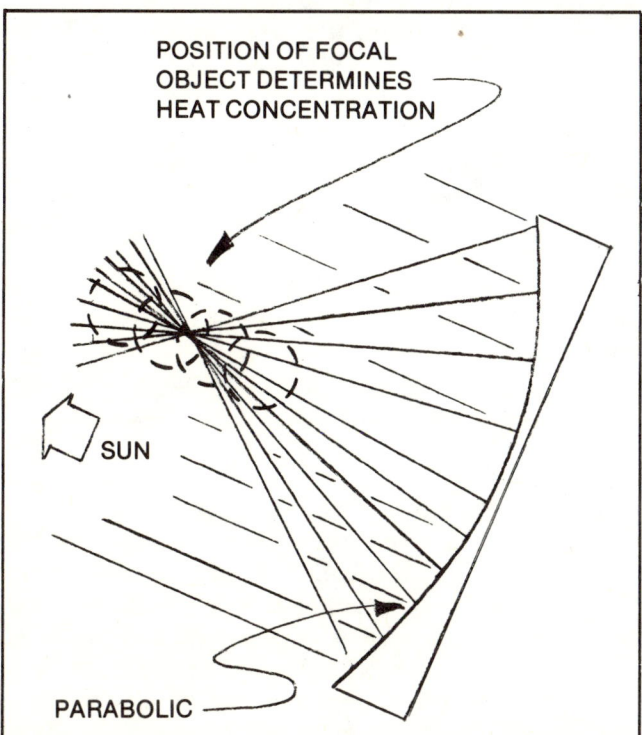

Fig. 2–19. Shifting the focal object relative to the focal point

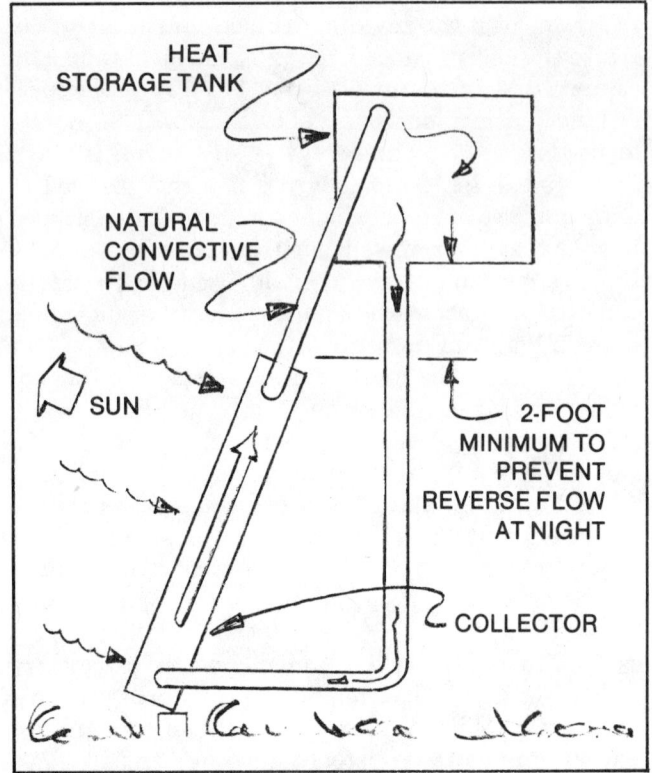

Fig. 2-20. A thermosiphon system

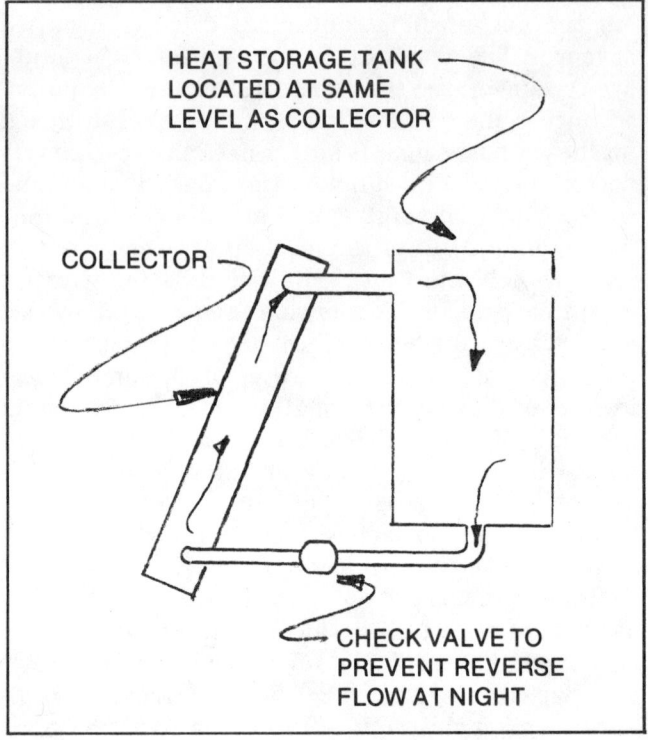

Fig. 2-21. Using a check valve in the thermosiphon setup

thing takes its place; that's the cooler medium. Arrange it correctly, and you will get a siphoning action, hence the name "thermosiphon."

The thermosiphon process abides by some strict conditions; if they don't exist, it doesn't. Take, for instance, the size or type of tubing, if used. If it's small, say 1/2-inch or less in diameter, the resistance to movement of water or glycol is too high and it won't work well, or at all. We may be dealing with a matter of degree. Since heat is leaky stuff, we will want to keep our collector operating at low temperatures. Thermosiphoning will eventually occur in even the most restrictive system, but at so high a temperature differential that our losses would reach intolerable levels.

Other things affect thermosiphon action. An arrangement of tubes, as illustrated in Fig. 2-39C will not allow thermosiphoning; the heated medium in each tube will be "bucking" the action of other tubes as each tries to rise. The layout shown in Fig. 2-39B will probably thermosiphon, but only with some difficulty and certainly at too high a temperature for the collector to be really efficient. Thermosiphoning depends on a closed loop of medium, meaning both a continuous pathway and one completely filled with medium. The slightest air bubble can stop the thermosiphoning, building enormous pressures and temperatures as the collector system fails to rid itself of the acquired heat. A means of bleeding bubbled air

from the pipes after the initial installation (at the highest point, or points) is required; these are usually the fill valves used to put the medium in the system. Temperature-and-pressure (T&P) valves are also sited there, for protection in the event the system leaks, air bubbles form, or all air was not bled from the system. Thermosiphon systems should have a medium which cannot be vaporized with high heat, lest this be the source of air bubbles. Periodic bleeding of the system will reveal if this is happening, or if the system is leaking somewhere (and letting air in during the process).

Traditionally, thermosiphon systems (with a water medium) have the bottom of the storage tank above the top of the collector (Fig. 2-20). The storage tank can be positioned lower, but a nasty thing happens at night: the system can reverse its flow. Since there will always be some warm water in the collector at the end of the day, it gets cooled during the night, settling downward just as the daytime-heated water wants to rise. It then pulls heated water from the storage tank into the collector, cooling it. After a while, a regular flow starts and the day's precious labor is wasted. Instead of the shower's good morning warmup, we get icy sleet.

To prevent this, storage tanks are placed above the level of the collector. This isn't always practical or desirable; if the collector is on the peak of the roof, where are you going to put the tank? But if the storage

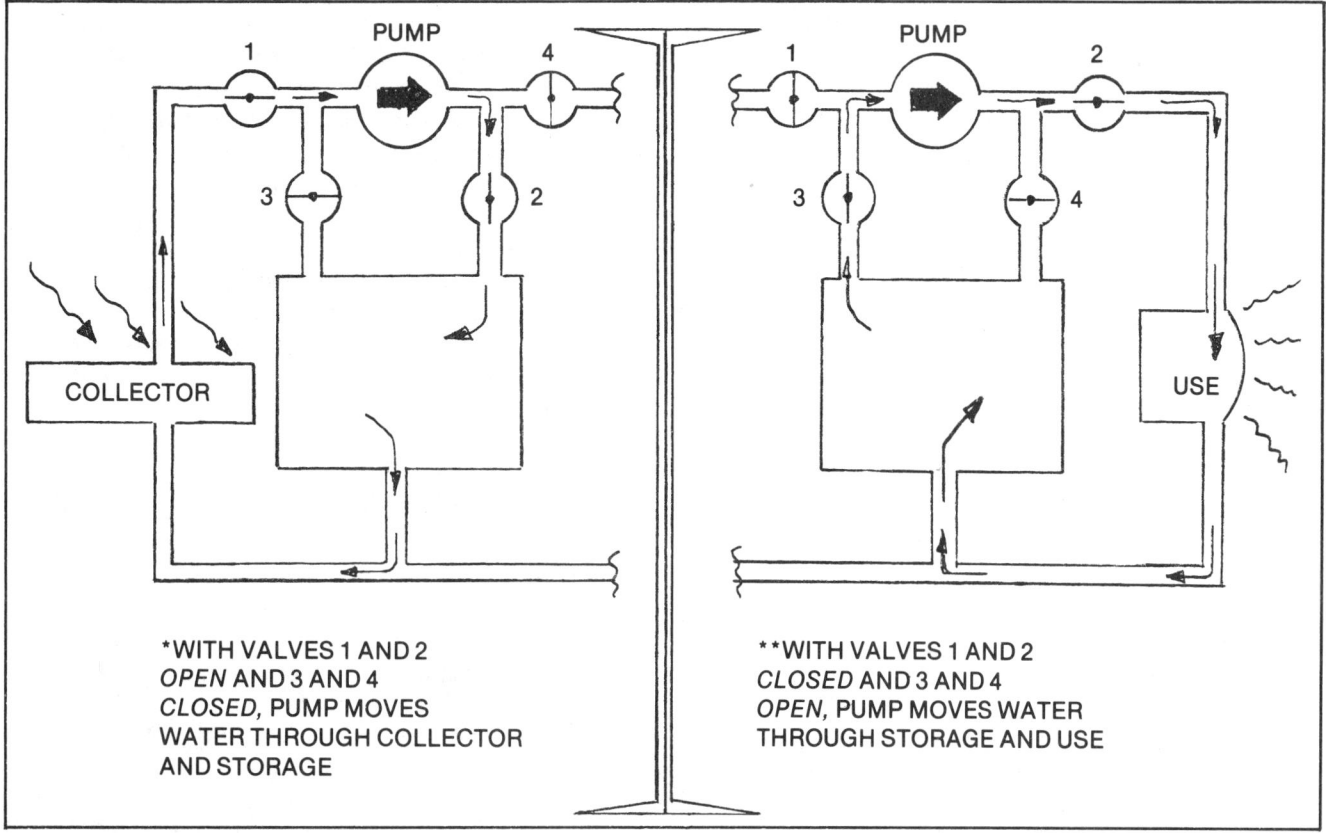

Fig. 2-22. A double-purpose pump

tank is on the same level as the collector or below it, a check valve must be used somewhere in the system, usually in the cold-water feed tube to the bottom of the collector (Fig. 2-21). There are number of valves available, but the preferred one is the gravity type; gravity closes the valve and any backpressure holds it closed. Be careful that it is inserted in the correct direction, and right side up. Avoid the spring-loaded type of check valve; they're fine for pressure systems, but thermosiphoning is a wafting type of pressure and it won't have the strength to open the valve from either direction.

Just how far you can put the storage tank below the collector array depends on many factors—size of the system, pipe size and lengths, etc.—and there's no rule of thumb. To operate the collector efficiently means a minimum temperature differential between what's heated and what hasn't yet been heated, and this isn't going to make for a strong thermosiphon action, severely limiting how far you can locate the tank below the collector array. The safest situation is locating the tank above the collectors, making a check valve unnecessary. The next best bet is to put the tank directly behind the collector, at the same level, and use the check valve. Anything else, and you're on your own. Calculate the height you can get away with or push your luck. Or pump it.

PUMPED WATER

If thermosiphoning won't work for the type of collector or system you're using, or you've other reasons for wanting a pumped system (unbeknownst to me), then by all means pump it. This isn't as involved as it may sound. A household-sized system uses a pretty small pump. It won't have to pump very many gallons per minute, and it won't have to pump very fast; we're talking about something between 1/30 and 1/6 horsepower. If this pump is being used to deposit the acquired heat in storage, and you won't need the heat until after the day is almost over (when the pump will not be in use anyway), you can, through some fancy design work, make the pump work for you in moving the heat from storage to use (Fig. 2-22). The reason is to eliminate waste motion, not expense. Too often, if you're thinking of saving bucks, you've compromised the system.

With a pumped system, you can put the tank just about anywhere you want—in the basement, down the block, etc.—because you're supplementing the thermosiphon action, or replacing it. Depending on the pump, you may still need a check valve; just because the pump is stopped doesn't mean water (or whatever) can't or won't flow through it. Check with the manufacturer or check the pump specifications before you assume either way and pick a check valve that activates well below normal pump pressure.

The pumped system, for all of its positive action, has one powerful negative factor. What happens if there's an electrical blackout about high noon in the middle of a July day? That's what the T&P (temperature-and-pressure) valve is there for, but the system is going to end up dumping a lot of water or antifreeze, and you're not going to have a hot shower when you get home from work. Give it some thought energy now. Reliability is a great virtue in your solar energy system. And it won't be with the increased utility blackouts we'll see as consumption confronts diminishing resources. Of course, it's a different matter if you've got a wind-electric system and the batteries operate the pump directly. Or, if you can't (or won't) abandon the utilities, you might examine low-voltage pumps and use the utility power to keep the battery charged that runs the pump. This operates on the same principle as the emergency lights in schools, hospitals, etc. that go on when the breakers trip out. Or use a thermosiphon system.

WATER PUMP/SIPHON

Better still, use both thermosiphoning and a pump. While the former can't really back up the latter, a pump can back up a thermosiphon system. With thermistors (resistors that change valves when heated or cooled) embedded in each collector array, a blocked-flow condition caused by trapped air will become immediately evident; that collector will get hotter than normal. This can activate a relay to turn the pump on for a few seconds up to a few minutes (any predetermined time) to wash out the air bubble. To work, the pump must be installed in parallel with the system's check valve (see Fig. 2–23), thus assuring a

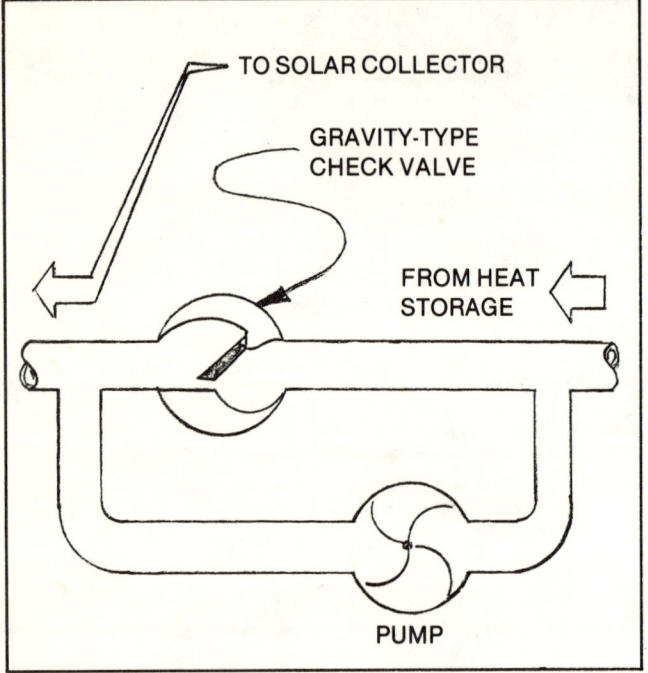

Fig. 2–23. A pump by-pass for the thermosiphon

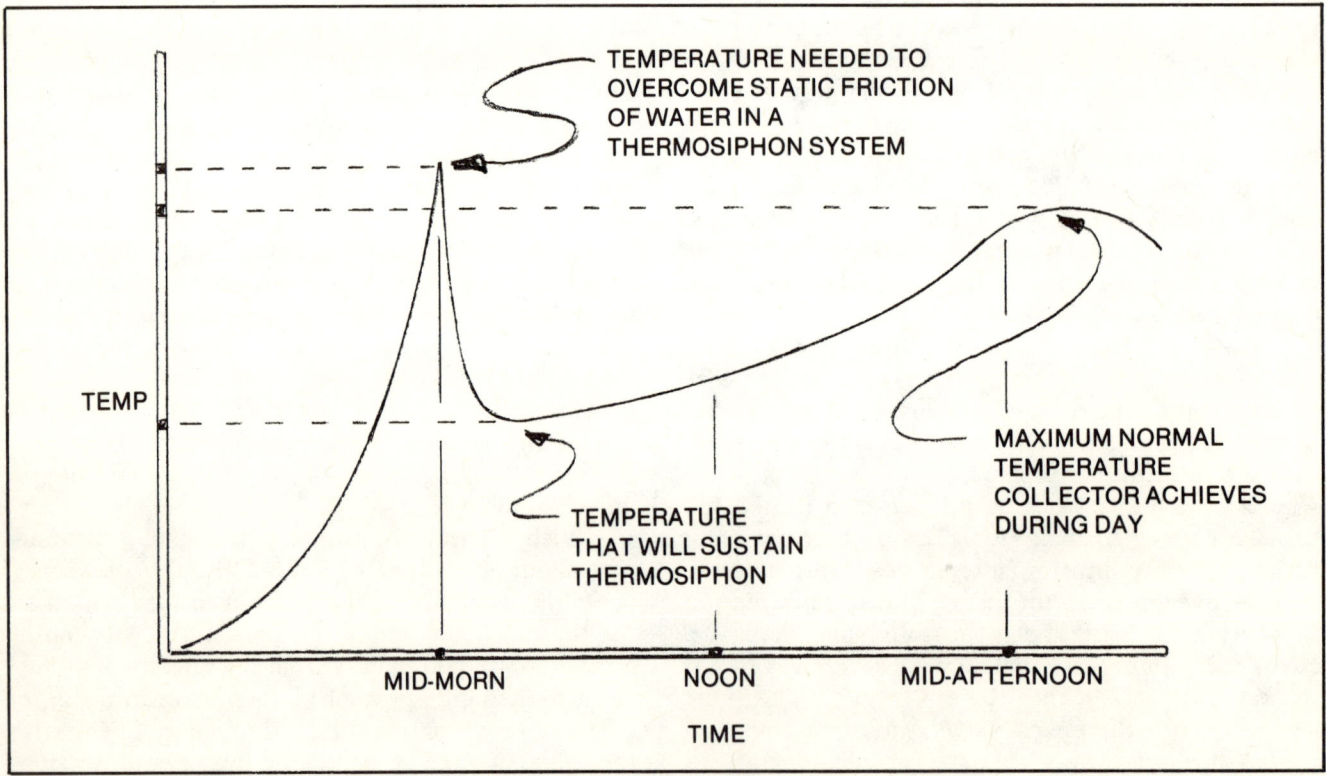

Fig. 2–24. Daytime temperatures in the thermosiphon system

good passage for thermosiphoned flow when the pump is not on.

Another useful purpose for a paralleled pump in a system designed for thermosiphoning is to kick the system into action initially. As illustrated in Fig. 2-24, the temperature attained in the collector prior to the first bit of flow at the start of day is much higher than the one which will sustain thermosiphon flow. This shouldn't be too surprising; many things go easier once they're started—motors, bicycles, writing books, etc. The trouble with allowing this to happen as it will is that, once the flow has indeed started, cool water flows into a relatively hot collector, giving it a "thermal shock." This is a rapid cooling or heating, which causes rapid contraction or expansion. Even if we are dealing with the same materials—and most often we're not—they can expand or contract unevenly because of thickness or position; if we've got a thermal bonding material between them, we may lose part, or all, of that good conductivity. Maybe not the first time, and maybe not the hundredth, but sometime during the lifespan of the collector, the thermal-bond cement may come unglued. The pump is set to turn on, as we know from our previous discussion, whenever the temperature rises above normal operating level (determined by empirical observation). It should, therefore, turn on when the temperature rises above normal for initial flow, kicking the system into flow, and turn off to permit thermosiphoning. It's the way to start the day, right?

AIR-MEDIUM THERMOSIPHONING

Air as a heat-transfer medium poses little difficulty compared with water-medium systems. With a density that's 1/800th that of water, air moves if you just look at it the right way, so thermosiphoning is a cinch. But what applies to water applies to air when it comes to the location of storage. If storage is placed above the collectors, everything's ducky. Put it directly behind or below the level of the collectors and at night we'll get reverse flow in the system and give away all that hard-earned heat (without a check valve of some sort). Air, as well, moves around things in much the same way as water, so ducting bends must be wide and smooth. If several collectors are connected in series or parallel, the ducting from each must terminate in proportionally larger ducting running to and from storage.

Many systems using an air medium have blowers to move the heated air to and from storage. Although it decreases the versatility of the system, ingenious designs may use the same blower for both functions. However, the more common installation uses two blowers—one for each direction—to accommodate the slower rate of flow in the collector.

PRIMARY VS. SECONDARY SYSTEMS

In the basic solar water-heating system, the water which moves through the collector is the same water that comes out of the faucet or the shower head (Fig. 2-25). This is called a primary system. It works fine in Southern California, Hawaii, and the South Sea Islands. Elsewhere, your collector pipes burst with the first real freeze. Or you must dump the water from the collectors and go to an alternate water-heating system (you did install one, didn't you?). Or, use another type of hookup, called a secondary system (Fig. 2-26), in which whatever flows in the collector as a heat-transfer medium is not water. Often, it's glycol, which we know best as antifreeze, the stuff that goes in car radiators. Put it, or a mixture of water and glycol, in the collector and there's no further threat of pipes bursting. We wouldn't want to drink

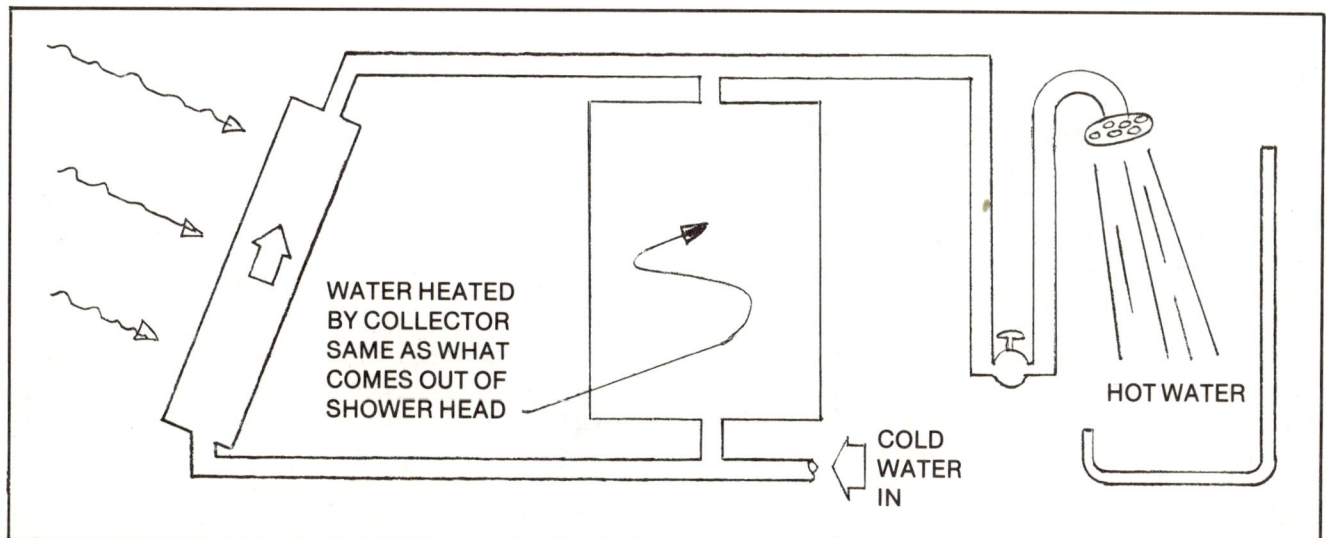

Fig. 2-25. **A primary water system**

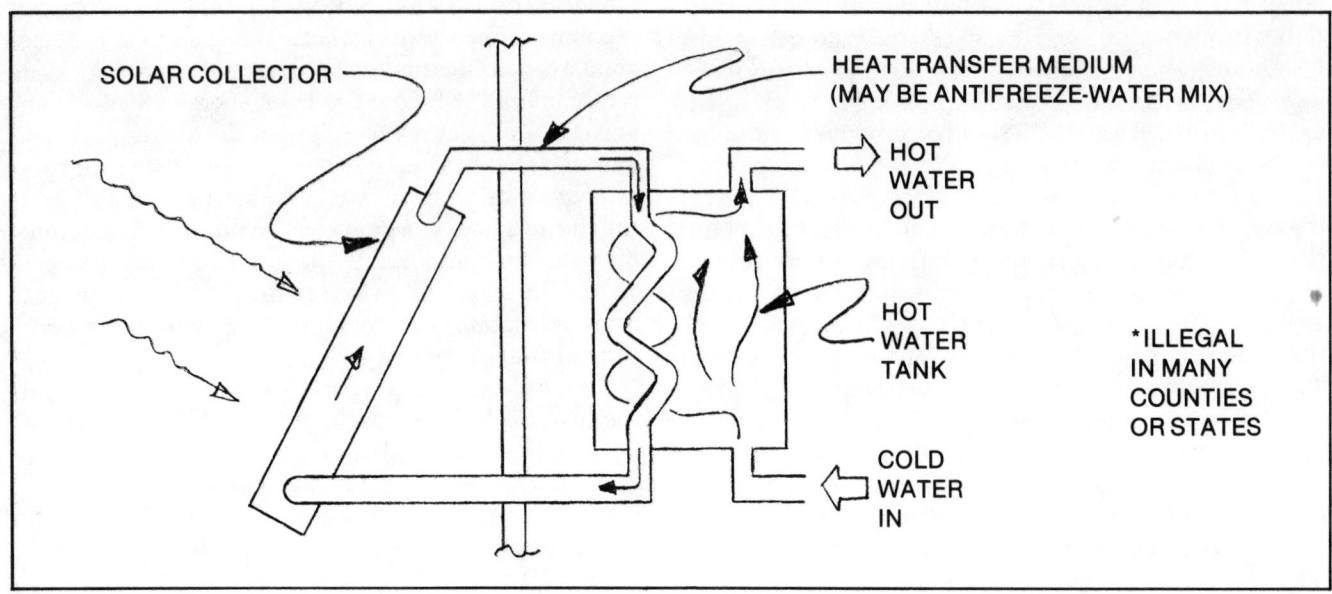

Fig. 2-26. A single heat-exchange (secondary) system

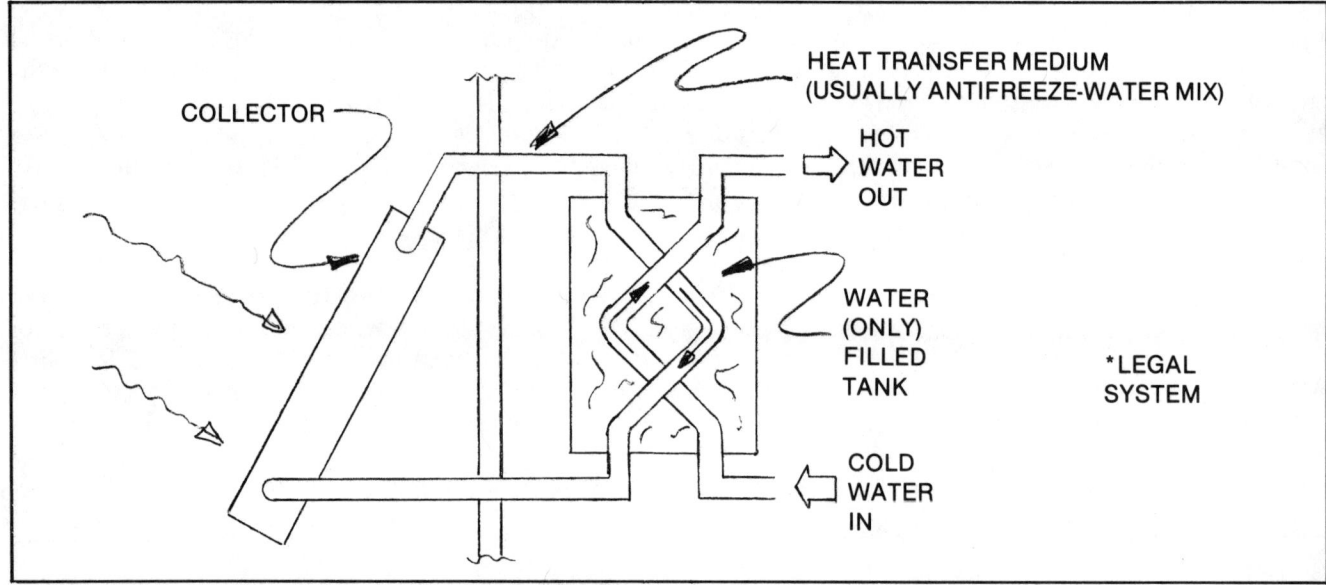

Fig. 2-27. A double heat-exchange (secondary) system

this stuff or take a shower in it, though; even if it wasn't deadly poison, it'd be an expensive proposition at even 10 gallons a day, much less 100! So, what we do is let it run through the collector and storage tank, and we allow the water we use to run through that tank also. If one transfers the heat to the other, we have a heat exchanger. It's cheaper to have the tank hold the water and have the coil of tubing carrying the heated glycol run through it; the tank's insulated, so it will not freeze. In practice, building codes specify that the heat exchange must be a double one, as shown in Fig. 2-27. Most antifreeze is not only deadly poison, it's tasteless. So you wouldn't know if some of it leaked out of the tubes into the drinking/cooking/bathing water. In a double system—with the tank water isolated from both the collector solution and the household water supply—a leak in either tube will not kill you or a loved one. Cheap insurance, isn't it?

A secondary heat-exchange system boasts a few worthwhile by-products which may even be of interest to those who could use a primary system. One is that antifreeze is slicker, lubricating its own passage through pipes and collector. It doesn't boil and expand as water will, a nice asset which keeps it from building up pressure in the system under heat duress. Even if water is used in the collector-storage loop of the secondary system, it will be relatively free of bubbles and, if distilled, will prevent the corrosion, sedimentation, or deposits that city or well water imposes on the innards of the collector and connecting pipes. And finally, if your pocketbook is perpetually empty, you may use materials in the collector that could not stand the corrosive effects of water, such as steel tubing; glycol is rust-inhibiting.

The primary system is inherently more efficient because it has less transferring to do, so it's not a clear-cut case for the secondary system; you'll have to weigh the factors if you live in a nonfreeze area. The rest of us have the decision made for us, and can happily live with it.

DUCTING/PIPING

Since the heat-transfer medium can't teleport itself from collector to storage, we must connect the two with pipes (for fluid) or ducts (for air). Both require insulation and, where exposed to weather, waterproofing of the insulation. Interconnecting tubes should meet, or exceed, the diameter of tubing used in the collector, and wherever they are common to several collectors, the inlets and outlets should be sized to accommodate combined flow rates. Outlet pipes should be inclined slightly upward and never run perfectly horizontal or dip below the horizontal; this assures that the flow of heat-transfer medium cannot be blocked by trapped air. Inlet pipes should be inclined downward toward the collector for the same reason. Use nothing sharper than a 45° bend in the interconnecting tubes to minimize pipe resistance to flow. Plastic pipe, if there are no other objections to its use, should be considered before galvanized pipe for the smoothness of its interior walls.

STORAGE

Squirrels put away nuts for when no nuts will be available, just as we must put away heat for when it is cold. If we could store away all the excess heat from summer, we'd have a dandy winter. But heat is an escape artist and will thwart our every effort to store it. It is a battle that we will always lose, and this denies us the capacity to store heat for many, many months. With careful design, good techniques, and sparing use, however, we can store it on an hourly, daily, and (sometimes) even weekly basis, using what energy we need as we collect it and squirreling away the rest. How can we do this inexpensively and effectively?

After going through the long list of what can store heat, we're left with only three practical candidates—water, rock, and salts. Each will store a quantity of heat; how much per pound or gallon of each is the question. To assess them comparatively, we must describe them in the same terms; one way is by "specific heat."

SPECIFIC HEAT

The Btu (British thermal unit) is a measurement of heat; think of it as a little bundle of energy. Just one of these will increase the temperature of 1 pound of water by 1°F. Two Btu's will raise it 2°, or 2 pounds 1°. Ten Btu's will increase the temperature of 5 pounds 2° or 2 pounds 5°. Everything is compared with water; since it takes only 1 Btu, it is said to have a specific heat of 1. Rock, on the other hand, has a specific heat of .24, which means that it takes only .24 Btu to get a 1°F change in 1 pound of rock. Or, 1 Btu will give a 1°F increase in 4 pounds of rock, or a 4°F increase in 1 pound of rock. Make sense?

Now, all of this holds true until certain points are reached. If water is boiling, or at the freezing point, several odd things occur. Say we cool a pound of water from 40°F. Just as you might expect, for each Btu that we take away from the water, the temperature drops 1°F. Extract 8 Btu's and the water drops 8° in temperature; now it's at 32°F. Extract another Btu and you'll see some ice crystals forming, but the temperature still reads 32°F. Extract another and another. Something's going on here. Ice crystals are forming, but the temperature still holds at 32°F. If we were to extract a total of 144 Btu's, we'd finally see a lump of ice and, with one more Btu removed, we'd see the first decrease in temperature, down to 31°F. The magic number of 144 Btu's holds true only for water; other liquids that crystallize into a solid at some temperature will have their own "latent heat" or "heat of fusion." These are change-state points and, when we're there, a whole bunch of Btu's need to be pulled out (or put in, in the case of melting) to effect a 1°F change.

So, what does this have to do with rock or water? Nothing, actually. It was just my way of introducing the third runner-up—salts. But now I'm getting ahead of myself. Let's start with water, go on to rock, and finish with salts for storage.

Water. Its abundance makes water a good choice for heat storage. Furthermore, it can double as the heat-transfer medium; rock doesn't flow through pipes very well (and it's awfully noisy). Water, with a specific heat of 1, can store 1 Btu per pound per degree (1 Btu/lb/1°F). Since there are a little more than 8 pounds of water in a gallon, that's a little more than 8 Btu/gal/1°F. If we raised the temperature of 100 gallons (two 55-gallon drums almost full) from 80° to 100°F, we would have stored away 8 Btu/gal/F° times 100 gallons times 20° temperature rise, or 16,100 Btu's. (for other conversions, down the line, 100 gallons of water, at 8 lb/gal, weighs 800 pounds).

The number-one drawback to using water is that it tends to be very leaky stuff, like heat. To contain it, we must have a tank. If it's made of steel, it usually ends up adding significantly to the overall weight of storage. Other materials are possible, using swimming pool construction techniques, but that's of little consolation if you wish to use a thermosiphon system and plan a rooftop installation. Who do you know who has a swimming pool in his or her attic? Lightweight plastic tanks or rubber liners offer the only hope to larger tanks. If you've got plenty of attic space, 55-gallon drums may help, but at 400+ pounds per drum, it's doubtful the ceiling joists can take it. The additional cost of a container for the water doesn't help much either.

Rock. Rock, like water, is an abundant material, making it also a good economical choice for storage of collected heat. Rock has a specific heat of .24, as mentioned, meaning that we'll need four times as many pounds of this stuff as water to store the same amount of heat. Well, that's the way it looks; let's check it. Eight hundred pounds of rock can store, with a 20° temperature change and .24 Btu/lb, 800 x .24 x 20° = 4,000 Btu's. Yep, that's one-fourth of 16,000 Btu's for 800 pounds of water. So, to store 16,000 Btu's we need 4 x 800 pounds, or 3,200 pounds of rock.

Now, we need more pounds of rock to equal water, but rock is about three times as dense as water. Unfortunately, a solid chunk of rock like that has very little surface area. When we talk storage, the amount of heat something contains is important, but even more so is the rate at which it will transfer the heat. Since that's a proportional relationship involving surface area, either we can't use the rock for storage or we're going to have to break it up into smaller pieces (2-4 inches in diameter) to effectively increase its surface area. That also means leaving air gaps in there, until, finally, we end up with a cubic foot of water and a cubic foot of rock having about the same density. Darn. Just when we thought we could get the same volume of each, that had to come along! So, it'll take four times the weight and four times the volume in rock to equal the heat-storage capacity of water.

Rock doesn't freeze as water can, however. And it won't corrode its container or leak out. It won't go in the attic, so you can't use thermosiphoning as a pumping technique if the collectors are located there. It helps if you're building your own house and can plan for rock storage, because it's terribly difficult to create new space under your present one. And even though rock doesn't leak, the heat it is designed to store will, so you must insulate it—top, bottom, and sides. Insulating underneath creates a host of problems, as you might well imagine; one of

Storage Medium	Btu's in 800 Lbs	Relative Capacity		Pounds of Storage Needed for 16,000 Btu's
Rock	4,000	1		3,200
Water	16,000	4	OR	800
Salts	96,000	24		133

Fig. 2–28. Storage capacity at a 20° temperature change (between 80–100°F).

them is how to keep the insulation there from being crushed paper-thin.

Salts. The eutectic salts represent the third means of storage of solar heat. This is not the common table variety of salt; eutectic salts are characterized by their low melting point. The most common or widely manufactured eutectic salt is Glauber's salt. At room temperature, it's a solid, just as ice is a solid at temperatures below 32°F. Raise the temperature of this salt to 88–90°F and it "melts." And, just as ice absorbs 144 Btu's of heat when it melts at 32°F, this salt absorbs a lot of heat—108 Btu's, to be exact—at this temperature. That's significant! Think of it. Raise the temperature of a pound of water from 88° to 90° and, with that 2° change, the water absorbs 2 Btu's. Now raise the same quantity of Glauber's salt (1 pound) the same 2°, and 108 Btu's are absorbed. That's what you call compact storage. And, you guessed it, the reverse is true as well. Drop the water 2°, you release 2 Btu's. The salt? A whopping 108 Btu's!

This has got to sound like the breakthrough of the century. And so it would seem, compared with water and rock. Glauber's salt has a specific heat of .7 and, for a 20° temperature change, that'd normally be only .7 times 20, or 14 Btu's per pound. But we have that 108-Btu gain because it "melts," which adds up to 122 Btu's! For the same weight or storage, 800 pounds, we'd have 122 x 800, or over 96,000 Btu's of storage. Let's bring the other figures to bear (see Fig. 2-28).

No matter how you slice it, a pound of Glauber's salt will replace 6 pounds of water and 24 pounds of rock. Or will it? As the storage temperature increases—a desirable situation for real life—water begins to catch up until, with a temperature increase of 80°, water requires only 50% more volume and, at 150° of temperature difference, they're identical. Why? The salts run into the same trouble as rock—they don't transfer heat readily and air spaces must be provided around the containers bearing the salts; water doesn't require these spaces and therefore makes better use of the volume.

All of the salts, including Glauber's, are plagued by other factors which, in the long run, prohibits their use. The salts are deadly poison; a burst container would be a major disaster. And speaking of containers, that's what's so expensive about using salts. The salts themselves are quite reasonable, but their containers aren't, costing ten to twenty times as much per pound as the caustic salts they house. Furthermore, field use of the salts has revealed a tendency of the chemicals in the salts to separate and stratify, impairing further solidification and consequently, their ability to release the heat they're storing. Newer "batches" may have solved this problem. Be careful of yard sales, though—you might get some of the old goop.

Water and rocks have it over the salts, unless money is no object. For retrofits, water beats out rocks and, for new installations, it's a toss-up or matter of prejudice. I like the idea of storage high in the building, and water, storing four times more than rocks per volume and weight, wins out. Besides, if the house catches fire, what good will it do to pull the plug on your attic's rock bin?

AIR-MEDIUM STORAGE

Serious use of air as a heat-transfer medium calls for more collector area and some type of storage for non-sun needs. It'd be nice if we could just fill a big space with all this hot air and then release it in prescribed doses for home heating. But it won't work. So, we must use the same storage media as for water heat-transfer systems—salts, rocks, or water.

Rock storage is a natural for air-medium heat storage. With all those gaps left between the rocks to increase the heat-transfer capability, it's a simple matter to "suck" collector-heated air through the rock bins. In the process, the heat is absorbed by the rocks and the cooled air is blown back to the collectors for a refill. Reclaiming the heat involves sucking home air through the rocks and blowing the warmed air back into the house. Two blowers, or one blower and some fancy ventwork, will do the job for collector-to-storage and storage-to-use heat transfer.

Water storage is also possible, although drawing the air over a huge tank won't work. Smaller tanks, barrels, or bottles (1-gallon size, plastic or glass milk-type) work better, exposing more surface area to the heat. Increasing or decreasing the storage size, as dictated by usage, is much easier with this system than with a basement or room full of rocks.

APPLICATION TEMPERATURES

Before we can happily collect and use the sun's energy we must compute, for each application, the final desirable temperature. Having a million Btu's of heat tucked away in your own personal storage system won't alleviate the bitterness of a tepid 65° shower, nor make it any hotter. Sure, you've kept the storage losses down with a lower temperature differential, but where's that hot shower? This is the snafu when it comes to dealing only with Btu's in heat that's converted and stored; we mustn't forget what we want to use it for. And, by inference, we shouldn't try to combine storage for different applications. The insulation alone required for a storage system that attempts to satisfy the requirements of both space heating and water heating would strain our sanity and pocketbook alike.

Section A of the Appendix: Data Cubbyhole details the computation of factors involving sizing of collector, storage, and use; I refer you there for the particulars. For the rest of us, a few generalities are in order which, I hope, will acquaint you with the futility of building a system that tries to do it all, all of the time.

SYSTEM SIZING

Storage must be sized for application and season. For any given application, too much of it means the quality or temperature is too low to bring us any happiness, as we discovered in the shower stall. Too little, and both storage and collector losses accelerate as their respective temperatures rise. So, you say, let's get it right. But for what season? The difference between the heat demand in winter and summer is obvious. So size it for winter. What part? Well, the worst part. But are you sure you want to do this?

There are peak demand periods in the winter, mostly confined to a couple months, but occasionally happening at odd times from freak storms, taxing our system to keep us happy. But let's look at the months on either side of the peak months. With storage designed for that differential temperature (collector and storage alike), we store away enough heat to compensate for losses. We're also concerned with a temperature high enough to be useful and yet low enough to avoid losses. Now comes that blast of real cold stuff, and your place is socked in for three days. Maximum differential temperature implies maximum losses. The system can't cope. You want to build it bigger—more collectors and more storage? How about 30% more? 60%? How about 100%? Still no? You've got to be kidding! Three times? Four? Ridiculous, isn't it? Some processes are linear and some aren't. And, unfortunately, this is the latter. A system that takes care of 100% of the demand may cost from four to ten times as much as one that takes care of 90%!

PASSIVE SPACE-HEATING SYSTEMS

There is another method of solar energy utilization that applies particularly to space heating. It is set apart from the systems discussed thus far by a num-

Fig. 2-29. Closing the lid on Steve Baer's drumwall to keep out summer daytime heat. *(Courtesy James L. Ruhle & Assoc.)*

ber of characteristics. One is that the solar energy collection, transfer, storage, and use blend into units which combine two or more of these functions at a time. Another is the bringing of these systems into the living areas; there are no collectors perched on the roof or storage tanks stuffed away into a cellar. All in all, the distinction is one of integration between home and system, and a genuine elimination of the bolt-on quality inherent with most installations.

No single term embodies the whole class of passive systems, but these three—the drumwall, the Trombe wall, and the Skytherm—come very close.

DRUMWALL

There is no denying the passivity of the drumwall. A stack of drums, filled with water and mounted horizontally on racks which will support the weight (at 440 pounds apiece) is placed along the side of a room or house that is covered with a glazing, say glass. The incoming solar energy passes through the glazing and is absorbed by the blackened drums, and the heat is transferred to the enclosed water. If the air in the room (or house) is cooler than the drums, the drums will radiate excess energy in the room. If the room is warmer, that heat will seek the cooler

Fig. 2-30. A drumwall interior

drums. To control the input of solar energy and the undesirable outflow of heat, some means of covering the glazing is crucial. This can be hinged panels on the outside, designed to lie down or open like garage doors in the daytime, or light-colored curtains or sliding panels of insulation on the inside.

This is a marvelous heating system. During the winter, the panels are opened during the day to admit sunlight. By day's end, we have a warm room and

Fig. 2-31. Water-filled columns absorb the sun's heat through the window; sliding door can close them off.

the tremendous reserve capacity of water at the same temperature. Btu's flow (by convection or radiation) from the water-filled drums into the room, replacing the Btu's that have found their freedom through walls, windows, and cracks. With the morning sun, the cycle begins again.

But wait! Wouldn't this system also cool the house in summer? Yes! Basically, the panels are opened at night, allowing the pent-up house and drum heat to radiate, convect, and conduct its way to the environment outside. Then, with first light, we close the panels. The cooled drums of water serve to soak up Btu's which find their way into the building and, without the input of direct solar rays, the interior stays cool.

I like this system. It's simple and effective. Steve Baer, the gentleperson who has used this system for a long time in the New Mexico desert, says its major shortcoming is the unknown life of the drums containing the water. But in every other way, it shines. Put together a framework. Set the drums in place. Fill them. Install the glazing. If it's glass, you can see out between the stacked drums and see people who can't see you. If it's a translucent glazing, the lighting it provides is gentle. Paint the drums your favorite color (the side toward the glazing should be black). If the movable panels are outside the glazing, you can cover the side that faces inward with a reflective or light-colored paint. This reflects additional solar energy onto the drums when open and, when closed, reflects heat radiated from the drums back into the room.

The purity of the system is retained if the insulating panels are mechanically raised and lowered, or swung open. With the noncirculation of the storage medium, this system could never be called anything but passive. Automating the opening or closing of the panels, by temperature sensing or time clocks, would make it an active system only if outside power was used. An extra of this system is its ability to use the Skylid (patent held by Steve Baer) to open and close louvers for the admission of solar light. This device used freon or other refrigerants which have a low boiling point. The Skylid's sensitivity to temperature differentials causes a flow of the refrigerant between two containers, upsetting the balance of a louver (to which they're attached) with the shifting mass. The louvers are "powered," therefore, by energy produced on site.

One variation of the drumwall is the water-column system. This is a series of tall columns of pipe or plastic filled with water; to maintain rigidity, they are usually of the corrugated or culvert type (spiral-

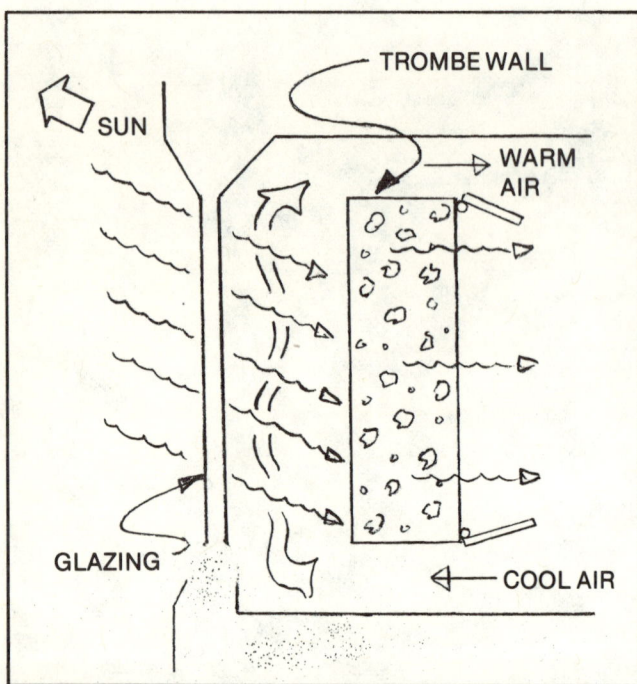

Fig. 2-32. The Trombe Wall system

Fig. 2-33. An adobe Trombe wall

ing). The effect is similar to the drumwall without the loss of interior floor area. When emptied, the columns can be moved about (within limitation) to best suit the needs or desires of the owner, and refilled.

TROMBE WALL

A concrete floor in any house can serve as heat storage if the sun-facing windows are large enough. However, the necessity of dark-colored floors to reduce glare and absorb enough energy has limited the use of this technique. A variation of this idea which eliminates the inherent disadvantages of the horizontal concrete floor is the vertical concrete slab, or Trombe wall. While this is similar to drum and water-column systems, it is different in several ways. First, it's not mobile. Second, to facilitate heat flow about the wall, it has both top and bottom vents (see Fig. 2-32). Also, an air space is preserved between the glazing and the concrete (or rock or brick) wall so that convection can occur. In the winter, cooler floor air is drawn into the apparatus and, once warmed, travels into the room via the top air duct; the circulating air gradually heats the room. At night, the shutters are closed and any heat retained in the concrete helps heat the room. In the summer, with shutters closed during the day, the only heat that gets in through the wall comes through the concrete; since the closed system rapidly heats up, the losses on the glazing side are high (which is desirable). At night, the shutters are opened and night sky radiation and other losses will cool the concrete and room air for another day.

Most Trombe wall installations are built without a means of controlling nighttime losses of heat in the winter and daytime "losses of coolness" in the summer. The shutters alone may work well in mild climates, but will prove insufficient in the more extreme. The lack of add-on or take-off storage means there's no room for error—you take what you've got, like it or not; everything must be calculated to the nth degree to avoid a horrible mess. But it *can* be done. If drums, water columns, and a lot of glass don't appeal to you, this is a real hides-it-all system.

SKYTHERM

Since heated air in a structure naturally rises to the top of the structure if allowed to do so, the Skytherm idea, credited to Harold Hay, is an excellent one. Bags of water are installed between the ceiling and movable roof panels over a house. When you want to gain heat, you slide open the panels and the blackened bags of water absorb it. At the end of the winter's day, you close the roof panels and the open the ceiling panels, letting this energy radiate to the interior. Or, if it shows a reluctance to do this, fans may be used to "destratify" room air, bringing warm air down to

Fig. 2-34. A Skytherm-type system open for the winter sun

floor level. In summer weather, the cooler water in the bags cools the heated air at the top of the house, which falls to floor level; the roof panels are opened at night to conduct, convect, and radiate away the collected heat of the day.

The Skytherm system works best in climates where the skies are clear, night and day, for long periods throughout the year—folks from California and Arizona, take note. If the area is also low in humidity during the summertime, a sprinkling of water outside and atop the water bags will further cool them by night-sky radiation and evaporative cooling.

Beside the limitations imposed by climate, a house using the Skytherm principle must be designed and constructed to withstand the weight of water involved, and fitted to drain off the water from a punctured water bag. Still, if you can buy water beds in bulk, this may be for you.

PASSIVE VS. ACTIVE

Don't judge the merits of passive and active space-heating systems by the number of pages I've devoted to each. Glazing requirements, heat exchange, storage considerations, construction techniques, etc. discussed for active space-heating apply to passive ones just as strictly. Owner-builders will find passive solar space heating much easier to apply to new construction than active systems. A retrofit for a finished home is a toss-up between active and passive installations; it may be easier to tear out a wall and go passive than to site the storage system and hang collectors on a roof that's oriented every which way but correctly. A most important consideration when selecting either system is compatibility with the existing structure. The house may look nice and the installation may look nice, but it's how they look and work together that determines whether or not you're going to have a successful marriage.

GETTING THE RAYS

The magic question is: How much of a solar collector will you need to do the things you want it to do? Unfortunately, that's an unanswerable question. If there is anything you learn from this next section, it should be not to ask it! If you really must, you deserve whatever answer you get.

There *is* a certain precisely measureable amount

of energy to be had from the sun's rays, and if you were in orbit, its value would be so constant that you could design your system to within a gnat's hair of what it would deliver. Down on earth, the story is entirely different. When we have absolutely everything almost perfect, we *might* get 300 Btu's per square foot of solar light. The only thing we know for sure is that at night we get zero Btu's per square foot. And, of course, we can expect to get something in between for the rest of the time. Kind of vague, isn't it?

To help you arrive at some average amount of energy you can expect to hit your collector array, I'll provide you with most of the factors that affect how much energy is going to get that far. Some of them you can do something about, but others of them you cannot. Since most of these factors vary, you will at least be able to assess when you might have to light off your backup system or break out the sweaters. Time of year, time of day, weather, atmospheric conditions, obstructions, geographic location, collector orientation, and collector types will all influence energy availability. Let's look them over.

Incidentally, some folks have gone to a lot of trouble to put together charts showing solar energy values throughout various parts of the world during each season of the year. These should be treated only as indicators, and you should be certain of what the numbers express—annual totals, annual average, season average, daily average for a specific date, or whatever.

TIME OF YEAR

It's fairly obvious to everyone that it's warmer in the summer and colder in the winter. You may even have heard that the days are shorter in the middle of winter than in the middle of summer. What you may *not* know is why. It's one of those things that you almost grasp, but it's hard to put into words. Seasonal shifts, wobbling orbits, being closer to and farther from the sun—these are but a few of the first attempts at explanation I sometimes hear. Knowing why it happens won't provide you with a means of changing it, and it's not going to make your collector work better, but knowing what does happen is *essential* to installing and operating your system correctly. The following discussion explains why we have summer and winter. Then we'll discuss the effect of that change and its importance to collector orientation. If you really want to get a firm idea of what the "angle of declination" is all about, I've worked out some examples in section B of the Appendix: Data Cubbyhole.

The earth is tilted on its axis at an angle of 23°. As you know, the earth makes a complete rotation on this axis in 24 hours, and a complete revolution around the sun in a year's time, or 365¼ days. If the earth weren't tilted, from any point on earth, at the same time each day, we'd see the sun in the same

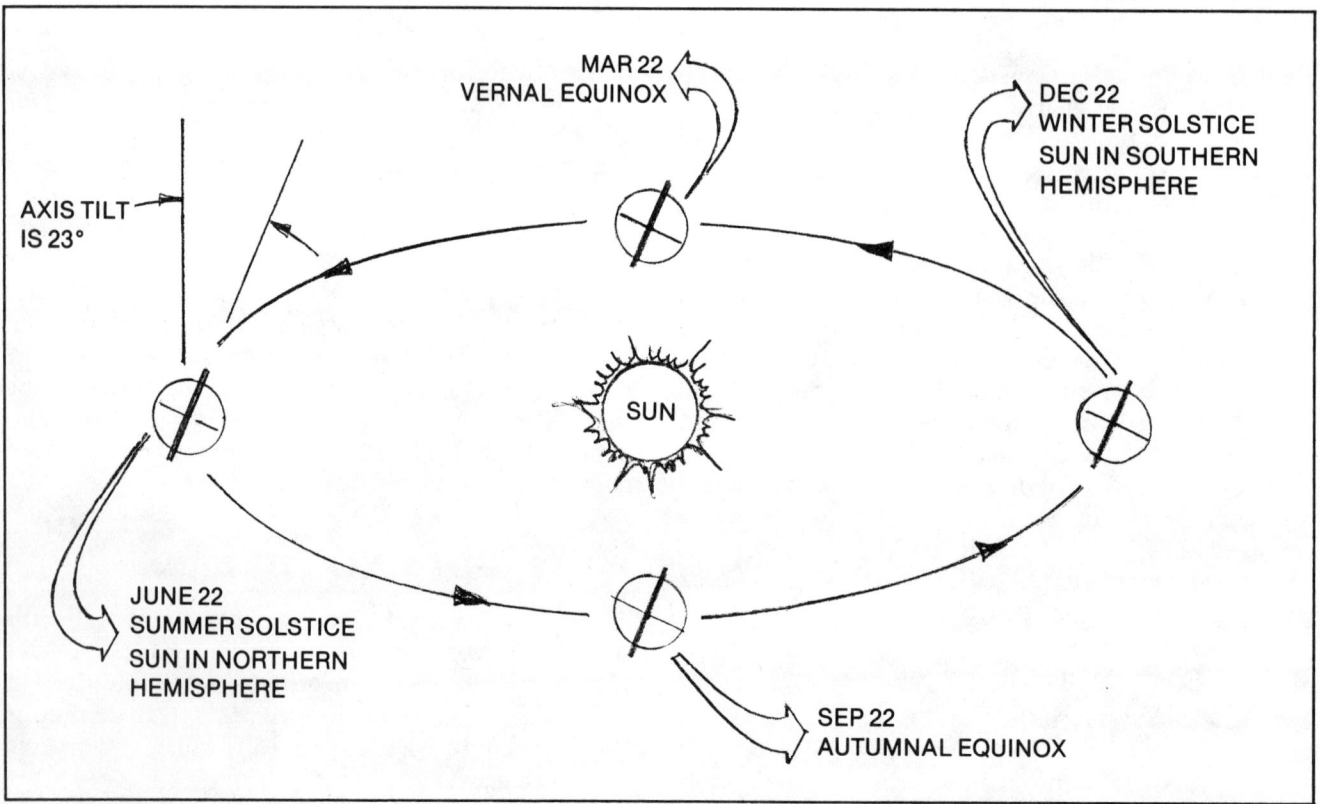

Fig. 2-35. The seasons

position in the sky. There would be no seasons, no need for daylight savings time, and life would be very dull. The earth's axis is tilted, but does not pivot, with respect to the sun. That is, winter in the northern hemisphere means that the north pole is tilted away from the sun, and summer means that the north pole is tilted toward the sun. Furthermore, this explains why there are six months of sun and then six months of darkness at the poles (study Fig. 2-35 closely). The two extremes mark the summer and winter solstice—the longest and shortest days of the year (and the highest and lowest positions of the sun in the sky) respectively. It's not easy to grasp the idea the first time around so I've provided some examples in Section B of the Appendix: Data Cubbyhole to cement the concept.

From the drawing, it should be evident that while there is only one of each of the two extremes each year, there are two times during each year when the sun crosses the equator; these are the vernal equinox (around March 21) and the autumnal equinox (on about September 23). Since the tilt is 23°, notice that the total swing from midsummer to midwinter is 46°. Back on the farm, from the farmer's point of view, it's obvious that the sun is the one that's moving around, highest in the southern sky in summer and lowest in winter. The maximum difference between these positions, then, is 46°. Aim your collector right at the sun in the summer, and you'll be mighty cold in the winter; with the change in the angle of declination, your collector will be misoriented.

Okay, that's twice that I've mentioned the angle of declination. What is it? Among other things, it's one method of describing the sun's position in the sky, relative to the point from which we're observing it. Of course, we can just look up and see it and say, "There it is!" Fat lot of good that's going to do for a blind person or a computer. Don't laugh—very few solar collectors are able to find it by themselves, either!

The angle of declination is determined by the time of year and your location's latitude. As the name implies, we're interested in the "declining" angle; that is, the angle the sun makes with a vertical line (or plumb bob). Even if you're reading this on June 21, you don't have to go out and measure it; it's easily computed. Simply subtract 23° from your latitude; the difference is the angle of declination for midsummer. Add 46° to this figure, and you have the angle of declination for midwinter. If your latitude is 37°, the midsummer angle is 14° (37° minus 23°), and the midwinter angle is 60°. Since we want the rays of the sun to strike the collector from the perpendicular, the collector is tilted up, off the horizontal, by an angle equal to that of the angle of declination (Fig. 2-36). Some systems have an adjusting mechanism for the collector for the angle of declination; if this is the case, a slight change very few weeks will maximize the amount of energy received throughout the year. This is the exception rather than the rule, however. If by decision or circumstance your collector cannot be adjusted, you will need to select an angle somewhere in the 46° sweep to optimize system efficiency. This is not as difficult to determine as it might first seem. The intended use of the energy will usually dictate this setting. A solar space-heating system will lean toward a winter setting, probably 10-15° from the lowest midwinter angle. A solar water-heating system is used evenly throughout the year, so it will be set only slightly favoring winter, perhaps at 20° from the winter angle.

TIME OF DAY

If your absorber-type collector is set at the correct angle of declination and set due south (for the northern hemisphere), the sun's rays will strike the collector glazing from the perpendicular at noon. Unless the collector is designed to follow the sun throughout the day, before or after this time it will strike the collector glazing at another angle. It's not surprising that the time of day will influence energy availability. This is due to the interplay of three factors—atmospheric length, reflectivity, and projected square footage.

Atmospheric length. At 12:00 noon, irrespective

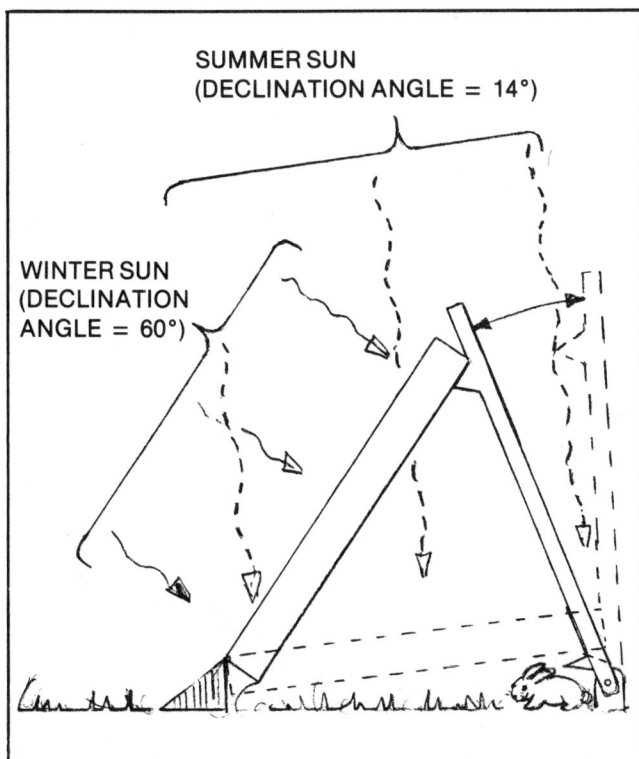

Fig. 2-36. Lowest and highest sun positions at latitude 37° North

of the angle of declination, the sun is at its highest in the sky for the day. It should make sense that at this time it has the least amount of atmosphere to go through to reach you, or your collector. If the sun is on the horizon (dusk or dawn), it will have more atmosphere to travel through, right? This turns out to be several times the amount it penetrates at noon. At that shorter angle, the sun's rays are also more prone to reflection off the atmosphere into space. At points approaching noon, on a curve between the extremes of dawn and dusk, the sun's energy is absorbed and reflected less and less by the atmosphere until we get up to the noontime maximum; the energy available to the collector follows a similar curve. The effects of dust, moisture, and smog in the atmosphere compound the problem, taking ragged chunks out of collector input.

Reflectivity. Only about 8% of the energy striking the collector glazing at high noon is reflected away, but this percentage increases with the hours away from 12:00 noon, whether morning or afternoon, as the angle of incidence increases. The reflectance losses are about 10% at one hour (15°, 11:00 a.m. or 1:00 p.m.), 14% at two hours (30°, 10:00 a.m. or 2:00 p.m.), 20% at three hours (45°, 9:00 a.m. or 3:00 p.m.), 28% at four hours (60°, 8:00 a.m. or 4:00 p.m.), and a shattering 55% at five hours (75°, 7:00 a.m. or 5:00 p.m.). These are optimistic figures; I've assumed that the glazing is free of dust and the collector is at the precise angle of declination for the time of year.

Projected area. With the sun beating down directly on the collector at noontime, we have an ideal situation—one square foot of sunshine hitting one square foot of collector. Again, however, the situation becomes less than ideal as we move away from the midday. If we would observe the collector at the crack of dawn or in the fading light of sunset, we'd see that *none* of the rays are hitting the glazing, since they're parallel to it. We'd still have a solid square foot of collector, and we'd have a weak but very definite square foot of solar light, but the two are operating in different planes, and we'd have zero interaction. It shouldn't be too difficult to realize that at points between sunset or dawn and midday, we'd have varying amounts of energy available to the collector because of this single fact: the projected square footage is not identical to the actual square footage (Fig. 2-37). This is the 6-foot man with a mile-long shadow effect.

How much more energy will ever make it to the collector? At noon, it is 100% of whatever gets through the clouds or atmosphere; this is called unity, or one for one. At 11:00 a.m. or 1:00 p.m., it's 90%. At 10:00 a.m. or 2:00 p.m., it's down to 87%. By 9:00 a.m. or 3:00 p.m., it's less than 71%. And at

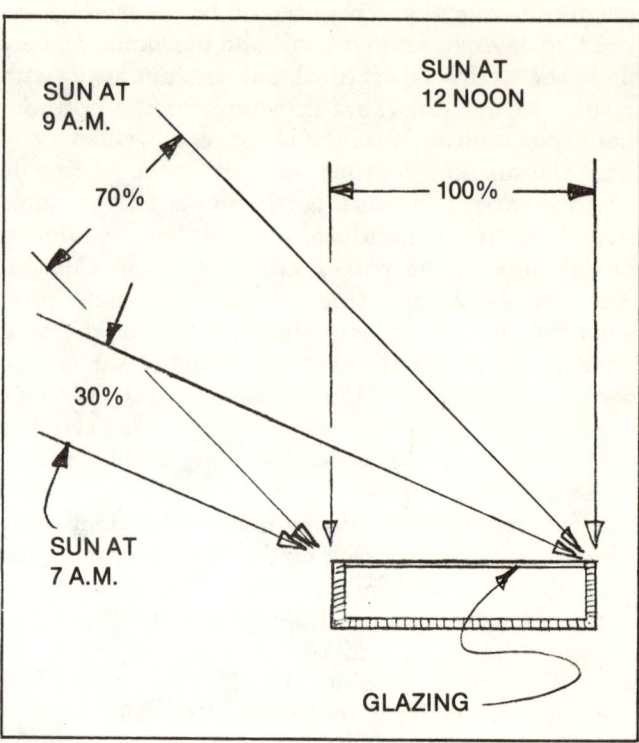

Fig. 2-37. The effect of "projected footage" at the collector

8:00 a.m. or 4:00 p.m., it's only 50%. Pretty sad.

The cumulative effect of just these three factors—atmospheric length, reflectivity, and projected square footage—is ample explanation of why solar collector design rarely considers more than a six-hour "working day," and never computes beyond an eight-hour day for nontracking absorber collectors. Even if we only considered a 5% loss due to atmospheric length, we end up with only 53% plus or minus at three hours, or a meager 34% at four hours, of the energy coming through at noontime.

OTHER FACTORS

There are a number of other factors which will affect the reception of solar energy at the collector. Three of the most important are geographic location, weather, and obstructions.

Where you live will certainly affect solar reception. Solar heating at the north pole will match the quest of a moth for a star, compared with even the northernmost portions of the U.S., so latitude is the first real factor. Trace any given latitude across the width of the U.S., check the weather data from these areas, and you will find a large variance in weather. Part of this can be explained by altitude; as rapidly as the atmosphere thins with altitude, it's little wonder that even a few thousand feet can mean the difference between a tan and a sunburn. Living in a deep valley at 10,000 feet will be of little consolation, however; mountains don't tend to transmit the sun's rays! Flat land doesn't necessarily guarantee good solar reception either, as residents of Los Angeles can attest; smog gets in the way.

Weather becomes a deciding factor in utilizing solar energy. The biggest culprit is the cold; while this doesn't interfere with actual reception, it sure makes hanging onto the received heat difficult. Snow, fog, and heavy overcast will shut down the best system and, in the case of snow, render the system inactive even after the storm is over. Another common mistake is to forget to wash off the collectors to remove accumulated dust.

When selecting the installment site, whether it be at roof or ground level, some compromises must be made. Trees, houses, or other obstructions will not always allow a clear shot at the sun throughout the day. There are two mistakes I've seen on a number of occasions. The first is installing the system in winter and forgetting that those trees will have light-eating leaves on them come spring and summer. The second is installing the system in summer and forgetting to look south as well as east and west for obstructions. The winter sun's path through the sky is very low. Don't forget that big 46° sweep!

TIME/HEAT LAG

Is the shortest day of the year, the winter solstice, also the coldest? Is high noon, winter or summer, the warmest part of the day? Is midnight the coldest time of the night? Is the heat of the summer greatest on the summer solstice, the longest day of the year? The answers to all these questions are the same. *No!*

When is the hottest day of summer? Some people would place it in the middle, or toward the end, of July, which is four to six weeks later than the solstice.

Most folks would pinpoint the coldest month as early February, the hottest part of the day around 2:00 in the afternoon, and 2:00 to 3:00 a.m., the coolest part of the night.

The reason for all this is heat lag. Effects take a little longer to manifest themselves than causes. And this is principally due to storage. The earth itself stores heat, as do the oceans. It takes time to process heat. Transfer is slow. That's why the pool water is still cold when the day is so hot, and wonderfully warm for a dip in the cool of the evening. This is not just a matter of comparative comfort on the part of the observer; the impartial thermometer is witness to it.

Users of solar heating, particularly passive heating systems, will notice this phenomenon and will want to use it to the best advantage. The storage medium soaks up the Btu's that have sneaked into the house all day long, keeping the interior pleasantly cool on the hottest day. The air temperature wants to shift (with some lag) closer to the outside air temperature, but the storage helps maintain that pleasant difference well into the night. You can think of it as taking the edge off the day and giving it to the night. If the nights are also relatively warm, the storage is usually set up to make the most of night sky radiation and what little cooling the night may provide, to be more effective come the day. In harsher climates, the tendency should be to burrow in farther, enlisting the aid of earth storage, which lags even more than system storage.

LOSSES AT THE COLLECTOR

Despite the taxes of nature and the man-made additives to the atmospheric soup, a small portion of the solar rays get through and impinge on the solar collector. Weary as they may be, and desperate as we might be for a solar shower after installing the collectors, all of the taxes have not yet been paid. We'll group the next chunk under collector losses. Understand where that energy leaks, however, and you can prevent or minimize it.

All right, we have impact. Most of the light goes through the glazing material into the collector, but let's look at what happens to the rest. Some is reflected—about 8% even at high noon—and, of course, it gets much worse as the angle of incidence increases; 40% reflection at 75°. And some of the energy is absorbed in the glazing or at the interface (if there's dust covering the collector)—how much depends on how thick the dust is and how many impurities exist in the glazing material.

For the energy which has run the gauntlet and found its way through the glazing, we have impact on collector walls and the absorber surface. If you use foil to line the walls, the energy will reflect and, with the exception of some that goes right back out through the glass, most will impact the absorber surface. Here the energy becomes longwave and it will reradiate, convect, and conduct its way around. That which is reradiated should *not* find it easy to pass through the glazing, although it can be absorbed by it and conducted to the outside of the collector. The warm absorber will conduct heat to the air inside the collector, and, if able to do so, this heat will convect to the top of the collector, conduct to the collector box material and through it, and escape outside. Heat will also be conducted through the absorber material and to the heat-transfer medium. Once heated to the point where thermosiphoning can occur, this will cause cooler water or air to enter the collector and provide more of a "heat sink."

To help you select or build and operate your collector, there are a few handy indicators of its proper operation you should know, and other details of design and construction that should be incorporated. Read on.

COLLECTOR CHARACTERISTICS

1. The collector should remain cool to the touch—glazing, sides, and back. This means that the energy going in is *not* coming back out. We could not afford the bulk or cost of the insulation required to keep heat in the collector for any period of time, so a cool collector is one which removes the heat as quickly as possible for use or storage. The only way this can be done is if the collector operates with the least amount of temperature rise possible. This does not mean that we can use a 5,000-gallon-a-minute pump and have an efficient collector system; moving heated water through a maze of pipe and whatever we're using for storage only exposes it to a much higher surface area through which it can dump heat. Rather, we are talking about temperature differentials (input compared with output water) of 15–35°F, diminishing as the water becomes hotter.

2. It's generally agreed that the collector should have a minimum of thermal mass (anything which absorbs heat in fair quantities). After all, if the heat isn't being transferred to the medium and then use or storage, it's not helping the system. There is some controversy on this point, because the presence of some thermal mass would "flywheel" the system through momentary blockage of the sun by clouds. However, there will always be some thermal mass in the collector—the absorber surface, the tubes that carry the medium, the glazing, the collector box, etc.—which is unavoidable. Just don't intentionally add bricks or rocks or heavy metals.

3. The absorber will need a coating to help it absorb the incoming solar energy. Black is the natural color for this sort of job, indicating by its color that it does not reflect any of the colors contained in white (some will even argue that black is not a color, but the absence of color). Don't forget, though, that we're talking about what's visible to us—our eyes. Whatever coating we put on the absorber surface will be "seeing" much wider ranges of frequency than we can see. Is black (to our eyes) still the best? Until a few years ago, it was. Now they've found that a particular shade of green is better for absorber use than black, although marginally so. Interestingly enough, it is leaf green!

The absorber coat—green or black—should have a flat finish rather than a glossy one. A roughened absorber surface is desirable; it increases the surface area, tends to trap received light, and lessens reflection of solar light coming in at an acute angle. It's preferable that the absorber material have this rough surface; flat black paint with a wrinkle finish is all right, but not nearly as efficient at reception.

A more modern way of treating the absorber surface for super-absorption properties is by using a selective surface, a twin-layered application that traps the incoming energy and inhibits reradiation of longwave energy. Unfortunately, the double or triple effectiveness is offset by an even higher cost, because applying this to a surface requires sophisticated techniques and exotic materials.

Since the absorber surface is likely to have a fairly large coefficient of expansion, make sure you have a good bond of paint to the absorber surface; the heat and the stretch will tend to crack and peel it. High-temperature paints, like the one used to coat barbecue grills, would be a good choice. Realize that most paints are nonconductors of heat. This tells us to keep our coats thin, and not to bond painted materials together.

4. Parts of the collector will need insulation. We're dealing with the glazing, sides, top, bottom, and back. Since heated air rises, the top of the collector may need more insulation, at least R-11 to R-19. Figerglass insulation is inexpensive enough, but bulky; 4 inches is needed for R-11, and 6 inches for R-19. Styrofoam would provide the required insulation without using up so much space. The depth of insulation inside the box along the sides, top, and bottom cuts into the space available for collecting energy. If it's a choice between a lot of insulation and limited collection area, and lots of area and little insulation, the latter wins out except in the more extreme climates. Collector-box material becomes a deciding factor, too; an aluminum box demands insulation, whereas a wood box lined with foil can get by without added insulation.

The back of the collector deserves more insulation, since it is exposed to more of the absorber area than the sides, top, or bottom. This will affect col-

Fig. 2-38. A copper riser/header solar collector under construction

lector depth; be sure to allow enough space for the minimum of insulation that goes here. Also, the absorber should have some means of restraint within the collector box that prevents it from resting on, and deriving support from, the insulation. Otherwise, the absorber will crush the insulation and reduce its effectiveness.

The glazing is the real culprit for heat losses in the collector—but, considering the contradiction of its dual purpose, passing the solar energy and preventing it from escaping, it doesn't do a bad job. For all its failings—weighty, costly, breakable, etc.—glass does it best, passing and holding a very slippery form of energy. The gap between glazing and absorber is critical. Make it more than 4 inches, and convective currents carry away too much heat. Close it down to less than 1 inch, and the air between conducts too much of the absorber's heat to the glazing and outside. A gap of 1–2 inches seems to be about right. Even with 4 inches of insulation behind the absorber surface, the collector box rarely needs to exceed 6–8 inches in depth.

5. The collector box demands a measure of attention, in design and construction. Besides the glazing, this is the only other thing people will see, so it should look good. The choice of materials used in the collector box will be strongly influenced, if you're constructing your own, by your skills and experience and by material workability. Buying or building, weatherability is the main concern; unlike the glazing, however, the box may be protected with paint. Waterproofing assumes priority in construction techniques; copious amounts of glue and sealant will be needed in every phase. There are two schools of thought on closure of the collector box. Some advise leaving access for correcting minor problems—leaks, cleaning, repairs, repainting the absorber, etc.—and others advise working out the bugs and then closing it permanently. An openable collector will resist waterproofing. Use screws, not nails, with collectors made of wood, and pop rivets for aluminum or other metals. Before lining the inside of the box with insulation or foil, glue down a layer of Mylar plastic or wax paper—overlapping the edges as side meets back, top meets side, etc.—as a vapor barrier.

6. While a perfectly airtight collector will minimize heat loss from the collector, it will, depending on how much air space you've left in there, experience a condensation problem. Water in the trapped air will condense on the glazing's cool surface and interfere with incoming energy and hence collector efficiency. To circumvent this, small holes may be drilled in the bottom of the collector (where rainwater will not enter) to permit "breathing." This will allow some heat to escape, but it's better to lose some than not have any. Besides draining condensed water, these holes will drain water that comes in from leaks that have gone undetected; of course, you won't know that the collector has failed the rain test, but at least you won't have sopping-wet insulation. The holes need not be large; three or four ¼-inch holes should do nicely. These, or a few larger holes, should have screening over them to prevent the collector from becoming home to bats, wasps, lizards, and other assorted creatures. Then you won't have to wonder what the smell is on the first hot day your collector is operating!

7. Wind can do more than just damage or stress the collector or its glazing; it also has a cooling effect. This is similar to the chill factor—the combined effect of wind and temperature—that humans experience. A rapid movement of air past the collector takes away the escaped heat faster. As if the collector could feel, it senses this and the heat losses increase. Situating the collector array to minimize or eliminate this effect is not always possible or practical. Fortunately, when we install the units for the least physical effect of wind, we decrease the cooling effect as well. But even where the wind's pressure is not a factor, limiting the cooling effect should be.

8. A collector that does not transfer the heat it collects can reach temperatures of 500°F. It might go higher, or not as high, depending on when the losses through the back, sides, and glazing finally equal the amount of incoming energy. Or when something explodes. Here's where your T&P (temperature and pressure) valve earns its keep. Of course, you'll lose all the hot water. Take some consolation in the fact that at least it will shut off when the pressure or temperature drops; a burst pipe will continue to fill everything up with water. But, of course, the normal operating temperature will never exceed 200°F and only rarely will it surpass 150°F. Take your time in sizing pipes, bleeding air from the system, and adding a pump backup for untimely blackouts. Melted or blown collectors are embarrassing.

TUBE-ON-SHEET PATTERNS

If you've decided on the tube-on-sheet collector, get ready for another decision—the pattern of the tubes. We've briefly covered the vertical tubes applied to corrugated sheet, but the need for a smooth flow of fluid in the collector warrants mention of other tube layouts. Briefly, we have the snake, snail, riser/header, and Hubbard squash.

Snake. Looking for all the world as if a great copper snake had found its way into your collector, this layout offers good coverage of the area receiving sunlight. It can be fashioned from one long section of copper (or aluminum) tubing using a tubing bender (Fig. 2–39A). A more squared-off layout (Fig. 2–39B) involves straight sections of copper tubing with nu-

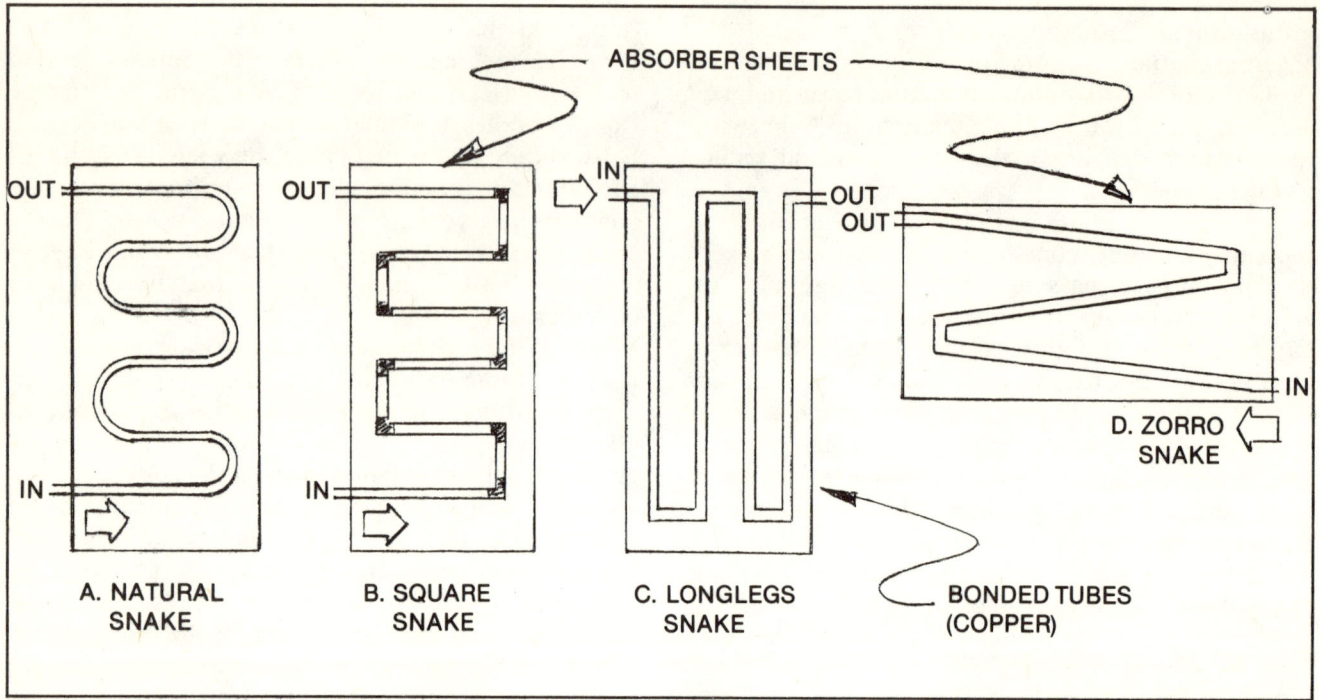

Fig. 2-39. "Snake" tube-on-sheet patterns

merous 90° copper elbows. To avoid this conspicuous consumption of wealth, we could go for longer sections of straight pipe and just a few 90° turns by arranging for a vertical layout (Fig. 2-39C) or a horizontal one (Fig. 2-39D).

Of the three, thermosiphoning will work best with A. The B diagram will offer a lot of resistance with those tight corners, but it will still work. Layout C will not thermosiphon. Actually, layout D would seem to be the best pattern of this type for thermosiphoning; without the horizontal sections, every portion of the tube helps the thermosiphon flow. As well, it is quite impossible for air bubbles to become trapped anywhere in this tubing. Unfortunately, this tubing arrangement does not optimize the collector area; in some spots, the absorbed heat will labor to reach the tubing and heat-transfer medium.

In a pumped system, layouts A and D will provide the least resistance to flow, because of their rounded corners, with B playing second best and C coming in last. If you have a very small pump and wish to keep it from "loading down," I'd suggest avoiding the C pattern. The thermosiphon process is not a designed flow technique—it occurs naturally. You may think of it, then, as assisting the pumped flow in patterns A, B, and D. Conversely, know that it resists pumped flow wherever the medium's motion is downward, as in parts of C.

Snail. In both layouts of the snail (Fig. 2-41, A and B), we see tubing spiraling inward upon itself in tighter and tighter curves, or, depending on your

Fig. 2-40. Flat-plate collectors using the "snake" design *(Courtesy James L. Ruhle & Assoc.)*

point of view, spiraling outward. Again, this can be fashioned with one piece of tubing (which would be interesting to watch) or made from straight sections with 90° elbows and consecutively shorter lengths of tubing. For either arrangement, I cannot imagine why anyone would want to punish themselves so, and yet I've seen this pattern again and again in working collectors. The only explanation that I can think of is that the owner-builder first experimented with solar energy by coiling up a hose in similar fashion. Then, noting how well it worked, this experimenter copied the collector pattern exactly. Obviously, there was a fear that the slightest modification would cause the collector to fail. In any case, thermosiphoning will not work and I pity the pump that feeds such a

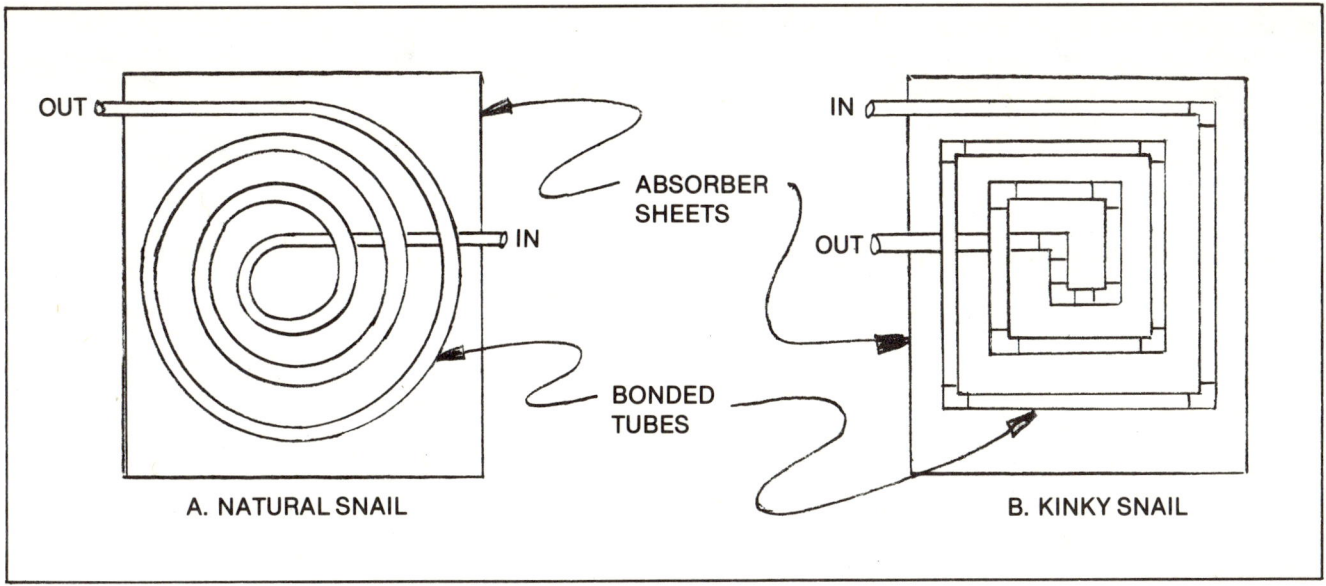

Fig. 2-41. "Snail" tube-on-sheet patterns

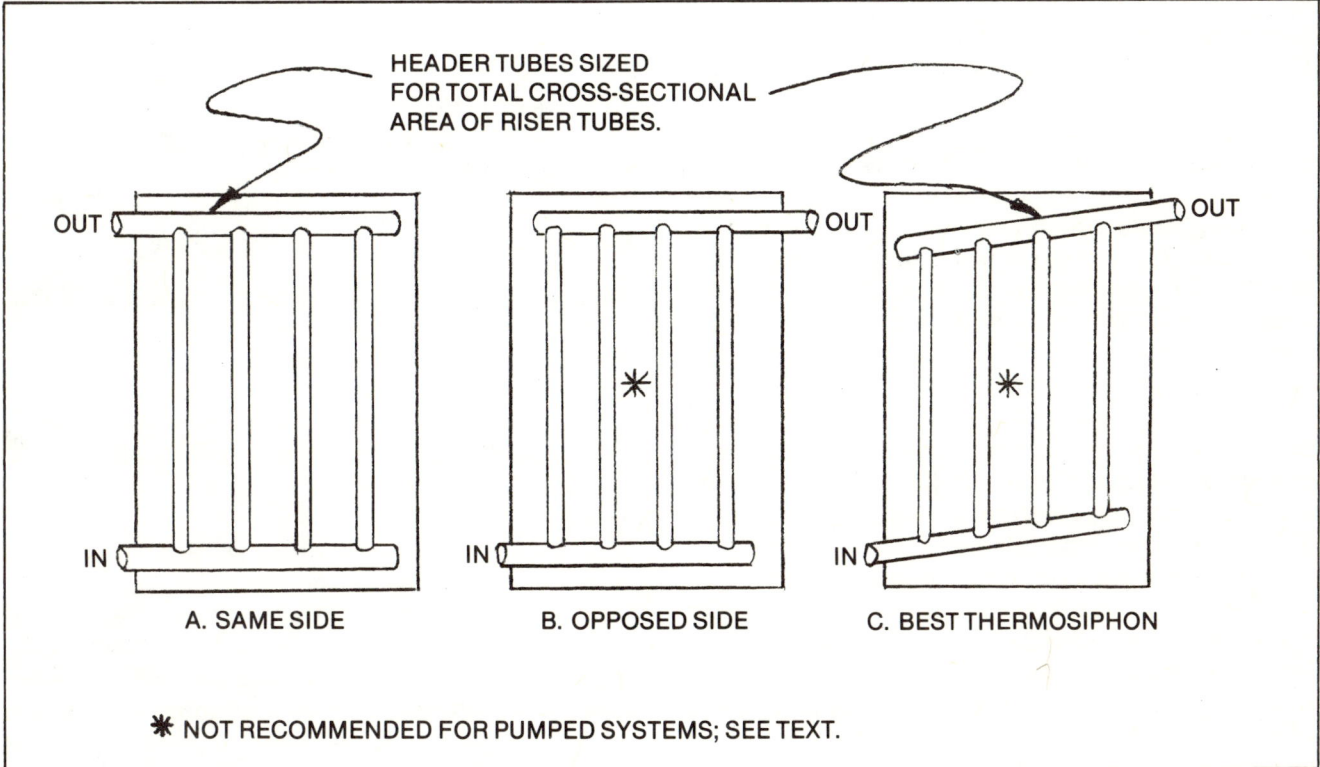

Fig. 2-42. "Riser/header" tube-on-sheet patterns

whirly-loop as this. And I don't envy the owner-builder who tries to bleed this arrangement of air when it comes time to add the heat-transfer medium!

Riser/Header. Parallel tubes aligned vertically, mounted on flat or corrugated sheet, and tied into larger-diameter tubes at the top and bottom form the third type of layout (Fig. 2-42A). The vertical tubes are called risers, and the larger end tubes are dubbed headers. The bottom header is the inlet and the top one is the outlet. The header tubes should be sized larger than those used in the risers, to accommodate the combined flow rates of all the risers. Typically, the risers are ½-inch to ¾-inch inside diameter (ID) while the headers are 1½-inches to 2-inches ID. These sizes are derived by adding together the cross-sectional area of each riser and finding a header diameter that will approximate this size. If you keep in mind that area is a square function of half the diameter, doubling the diameter gives four times the (cross-sectional) area. A 1-inch header, therefore, will handle the flow of four ½-inch tubes. It doesn't have to be an exact match to work, and won't be in practice—they don't make odd-size tubing. Just get it close or the header flow rate will be different than the risers; the resulting turbulence adversely affects thermosiphon or pumped flow.

A riser which is warmer to the touch than others alongside it indicates an interrupted or partially blocked flow in that tube. This is not always due to air bubbles or clogging debris. For some time, when the inlet and outlet of the headers were opposed (as in Fig. 2-42B), the third tube from the left (in a five- or six-riser array) would exhibit abnormal heating. Since this occurred in a pumped flow, it bore further investigation. Finally, it was determined that a venturi effect was actually causing a downward flow in the tubes on the left, no flow in the third tube, and varying degrees of flow in the others, with the highest flow in the extreme left and right tubes. To alleviate this, the outlet tube was brought off the same side as the inlet and the flow normalized.

Other than the Bernoulli effect of opposed headers, the riser/header pattern of tubing works equally well in thermosiphon and pumped systems. Since thermosiphoning doesn't suffer from the effects of a catty-cornered inlet/outlet setup, a parallelogrammed riser/header grid (Fig. 2-42C) would better serve thermosiphon flow than one with horizontal headers and eliminate any chance of trapped air bubbles. I could suggest that you merely tilt the finished collector, but it looks weird and you'll forever be explaining it to folks who are sure you couldn't hang a picture straight either. A compromise might be to install the grid at an angle within the collector box itself (Fig. 2-42C); the glazing should obscure the tilt.

Hubbard squash. The fourth type of tubing layout (Fig. 2-43) would appeal to the gardener for its resemblance to a Hubbard squash or a pumpkin. This type of layout makes fair use of the collector's area and is designed exclusively for thermosiphon systems. If you've just learned how to solder or heliarc-weld, this should challenge even the exceptional student. Actually, it's a nightmare of connecting tubes, and I've presented it because you might see it elsewhere and let its aesthetically pleasing lines overwhelm you into trying to build it. If it still does, go for it!

Bonding tubes to sheet. The selection of a tube-on-sheet absorber collector means that you'll be using some sort of thermal bonding cement, unless you're good at soldering and have selected materials that will take soldering. As with absorber paint, the

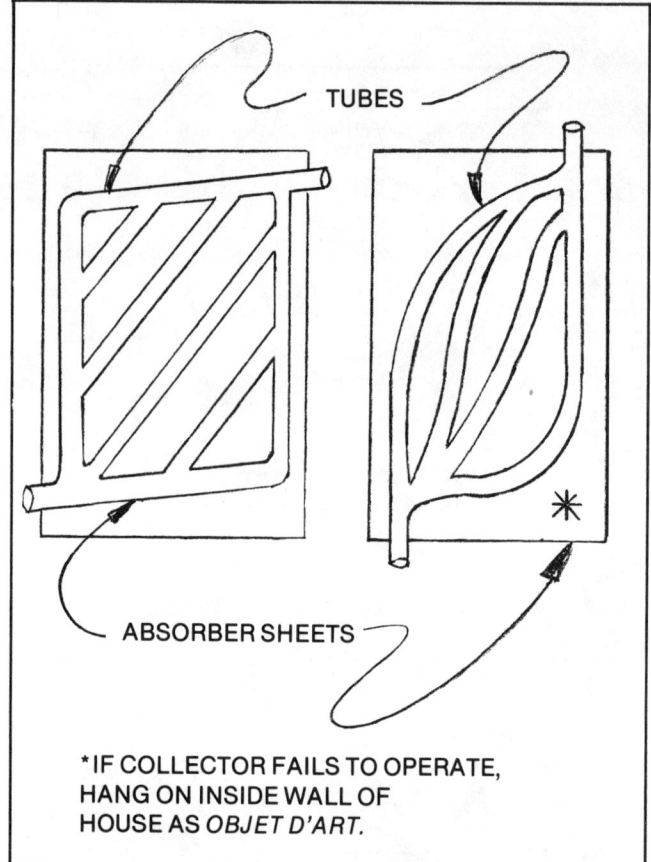

Fig. 2-43. "Squash" tube-on-sheet patterns

thermal bond of tube and sheet must be a good one, or we can expect it to pop away. When it has and it's obvious, we can curse a bit and try again. Most times, however, it won't be obvious; the tube will move only a fraction of an inch. So, it *looks* like it should work, but it doesn't. If you're troubleshooting a low-performing collector, don't forget to check this after ruling out other possibilities.

There are two ways to ensure a good bond between tube and sheet. First, roughen the surface of both the tube and sheet, just in the area you're going to bond, if you're doing it by hand methods (sandpaper, wire brush, etc.). If power tools are available, roughen everything; it'll absorb more energy and accept the absorber paint better too. When this is done, clean off the surface, using some lacquer thinner. This will remove the oil and dirt and then evaporate, leaving no filmy or oily residue. Apply the bonding material—more is better than less. I know it's expensive, but don't skimp.

The second way to obtain a tight bond is to connect the tube mechanically to the sheet. With holes drilled through the sheet at regular intervals on either side of where the tube will be bonded, use wires to tie the tube to the sheet. They exert pressure, keeping the tubes firmly against the sheet during drying. Furthermore, when you pick up the tube-on-sheet assembly, it invariably gets twisted and pulled, banged into things, or dropped in the process of getting it to and installing it in the collector box; the wire prevents cracking of the bond. After the collector is finished, the mechanical connection goes on to protect the thermal bond when the collector is raised and installed; to some extent, it protects from thermal shock during operation, also.

Whatever tiedown device is used, don't forget about dissimilar materials. If you've got copper sheet and tubes, copper wire will serve adequately. With an aluminum sheet and copper or steel tubes, keep the insulation on the wire when you secure the tubes, and prevent bends which will allow the rough edges around the drilled holes to slice through the insulation. Or, better yet, use plastic fasteners; not wire twist-ties, but captured plastic straps.

If the sheet and tube are dissimilar metals, the bonding agent should give a thermal bond without providing an electrical one. This is easily checked with an ohmmeter, or, if you don't know what that is, a small bulb and battery and three wires. Attach one wire between one pole of the battery and the sheet. Attach a second wire between the other battery pole and to one side of the bulb. Attach the third wire between the other side of bulb and the tube. If there is an electrical contact between tube and sheet, the bulb will glow; if not, it won't. Don't wait until you've got all the tubes on the sheet or you won't know, if the bulb glows, where the contact is being made. It's easier to find the unwanted contact if the tubes are to be connected to each other (via the header) after the bonding.

GLAZING REQUIREMENTS

When I've asked folks for the properties they would look for in a glazing, one item on the list is appearance, which, unfortunately, is usually the first item people consider when actually selecting their glazing. I mean, if you're buying from a company that sells glazing for solar collectors, won't they take care to supply all the desirable qualities? Well, yes and no. Yes, they try to meet most of the requirements. But no, it's not *their* responsibility to best suit *your* needs. Looks are important—for yourself and others who will see your installation—but a good-looking collector array is not necessarily a good operating system. A car may look elegant even when desperately needing an engine overhaul. Let's look at other glazing requirements. Check the following list item by item as you consider the glazing material.

1. Transmittance. The glazing should pass light. I mean useful frequencies of solar light; specifically, in the range of 0.1–2.0 microns and 3.5–5.0 microns. You should know the transmittance of the material you get, expressed as a percentage. Glass is usually high-transmittance, but large quantities of iron can ruin the effective transmittance. You can check the edge of the glass to see. A clear edge is good, but the greener it is, the more iron it has. Glass should not look smoky. Plastics will frequently not look clear; they may even look opaque. Don't judge by this. A collector needs frequencies that we can't see with our eyes. To those particular wavelengths, the glazing will be transparent.

2. Opaqueness to heat. Once absorbed, the solar light becomes longwave heat. We want to preserve this in the collector, preventing it from radiating back out. Glass and many solar glazings are opaque to this lower frequency and trap it in the collector—the so-called "greenhouse effect." But Mylar, for instance, will not trap heat in this way. Resistance to radiated heat should be as high as possible in our glazing material.

3. Weatherability. The glazing must stand up to every kind of weather experienced in your area. Realize that it must do this year after year, season after season. Will it hold up?

4. Ultraviolet resistance. Otherwise known as UV (pronounced you-vee) resistance, a necessary property in your non-glass glazing selection. Ultraviolet light does vicious things to plastic—discolors it, turns it opaque, dries it out, and cracks it. After a few months, you can begin to notice the effect. Without the UV inhibitors, no plastic really survives more than a year. If it says "UV-resistant," maybe it is and maybe it isn't. If it doesn't say that, you *know* it isn't.

5. Stiffness. Glazings that sag don't look good and don't work well, either. Call it stiffness or a low expansion coefficient—we want it for our collector glazing. Don't expect miracles here. You must support the glazing at least every 4 feet, if not at shorter intervals. But if you've done that, it should not sag when hot or, if applied hot (to get all the stretch out), it should not pop the fasteners when it cools. Follow the manufacturer's installation recommendations to the letter. It may be the only way your guarantee is good.

6. Thermal shock. If the collector rapidly cools or heats, the glazing must withstand this. I'm referring to pre-flow on a thermosiphon system and/or a blocked-flow condition before the alarms sound; melted glazing after a minor incident will ruin that evening chess game.

7. Cost. Usually the second criterion (after looks) for selecting a glazing, but it should be the last. That's because you want it to do just that—*last*. Cheapo materials won't. Okay, so you plan to replace it with something better later and, if you had to buy the high-priced spread, you couldn't operate the system this year with the present cash flow. You don't have to convince me; convince yourself. But look at the other criteria first, and ask the price last. That way at least you know what to get the next time around.

8. Impact resistance. If you live in an area that experiences hail, this will become a selective factor in glazing material. Glass is pretty tough, but it won't stop a golf ball or hard-thrown Frisbee, whereas some of the fiberglass-based glazings are shatterproof. Much may depend on where your collectors are located. For a ground-level installation, glass is a poor choice if whoever might be trying to catch the aforementioned Frisbee doesn't watch where they're running. Not even a shatterproof glazing can take that kind of punishment, but at least you won't be crying over spilled blood.

9. Collector size. Some glazings only come in specific widths (invariably about an inch short of your collector's width). Granted, you'll probably size the collector to the glazing or stay with normal widths, but this still will involve some preliminary checking. Just because plywood comes in 4-foot widths doesn't mean the glazing will.

10. Thickness. Give some attention to the thickness of the glass, if used as a glazing. Single thickness, or window pane, probably won't hold up to the assignment. Even if snowballs aren't part of your area's history, hard-driving rain, snow weight, and/or wind pressure can get it. It's a toss-up between double thickness and tempered glass; whichever has less iron, is least expensive, or is more abundant will do. Glass, like everything else, expands and contracts. Allow for this with a flexible-but-waterproof frame and seal, or use it in smaller sizes.

11. Other criteria. I can't think of any other important items, but that doesn't mean you won't; you know your situation better than I. Keep thinking. If nothing else, it's one way that people will know you're still alive.

INSULATION

At every phase of energy collection, transfer, storage, and use, insulation is needed. There are, of course, many losses that cannot be caught, so it must be our solemn oath to preserve what little we can. In the collector, insulation at the sides, top and bottom, back, and glazing performs a holding action on the received energy until it can be whisked away to storage or use. In storage, the insulation must preserve the precious heat from long-term leakage, from several hours to several days at a minimum. In use, insulation guards against further losses, preserving the heat in whatever application, for as long as required, desired, or possible.

Different materials serve as barriers to the move-

Material	Thickness in Inches	R Value
Brick	4	0.8
Cellulose fiber (loose)	1	4
Concrete	8	0.9
Cork	1	4
Mineral wool	3.5	11
Mineral wool	6	19
Gypsum board	1/2	0.5
Polystyrene	1	4
Polyurethane	1	6
Popcorn (in baggies)	2	4
Sawdust	1	2
Thermopane	—	1.5
Glass	1/10	0.9

Fig. 2-44. Material thermal properties

ment of heat. A way of comparing their effectiveness as an insulator is a must, and this leads to the evaluation of conductance. The test situation is a given area (such as 1 ft^2) of the material and a temperature 1°F higher on one side than on the other. What quantity of heat would flow at what rate through this barrier? Replace the test material with other materials and, when finished, you could make up a chart (see Fig. 2-44) that would reveal comparative findings. Those materials that did a good job of transferring heat would work well as conductors. The reciprocal of conductance is resistance. A poor conductor's reciprocal value would be large; if we place the number after R for resistance, we R-something. R-11 and R-36 are examples, with the latter being a higher resistance to heat flow. We aren't necessarily talking about two different materials, though; a larger thickness of the same material will obviously pass less heat in the same amount of time. Fiberglass, for instance, has an R-11 rating for 4 inches of thickness, and an R-19 value for 6 inches.

How much insulation do we need? Well, that depends. We're talking about two places: one is where you want to store the heat, the other is where it wants to go. The difference in temperature between these two places will affect the heat flow; the higher the differential temperature, the faster the flow. This means adding more of the same insulation (increasing the thickness) or using another type of insulation that does a better job for a given thickness.

The slowdown process of insulation is made more difficult by the different means heat uses to move about. Don't forget—it can radiate, conduct, or convect itself away. Materials which can stop one form of transfer can't necessarily stop another. Shiny aluminum is wonderful for turning radiated heat energy back to its source, but it is ineffective against, and may even aid, the escape of heat by conduction. A cellular material which hosts thousands of little air bubbles inside it, however, provides an excellent means of stopping conducted heat. Air convects heat easily, but it doesn't conduct it well; the dead air spaces, intermingled with the nonconducting materials, impose a large number of interfaces on boundaries, slowing down the otherwise smooth flow of heat through the material. It's similar to running a 100-yard dash—easier to do if you've got a dirt track than a slice of the Everglades.

Foil-backed fiberglass insulation retards all three forms of heat transfer, but since it's not rigid it can be used only where it won't be crushed—between floor and ceiling joists or wall studs. Blow-in or sheet insulation is also available; foil can be used with these to further decrease the escape of radiated energy. Other materials serve well as insulation, their use limited only by your imagination and the building codes—crumpled newspaper (or the pages of this book), sawdust, corncobs, cardboard, or old teddy bears. Maybe even popcorn; in times of depression, you could eat your walls. Eliminate or work carefully with any homegrown insulation which is combustible. Warm is fine; burned is not!

The earth is a wonderful insulator, if it is of the right type and depth. Dry, loose, and deep works best. It transfers heat, too, especially if it gets wet, but swings less, from day to night, season to season, than the air temperature. Cool houses in the summer and ones requiring little additional heat in the winter are a definite benefit. Maybe it's time to check into underground houses! A rash of new information is now available, and it dispels many of the myths and solves the few minor problems that have prevented widespread underground living. For those who are inclined to forego the serenity and quiet of living with mole and gopher, or who must contend with a house that's completed, an earth "bermed" heat-storage facility might be just the thing.

Simply burying the tank used for storing heated water in the ground won't work, however; the earth will rapidly draw off the heat until the tank achieves ground temperature. Insulation is still required, but once installed in the ground the storage tank will not have to deal with extremes of air temperature. This saves heat energy. Furthermore, it makes calculations of heat loss a lot easier, since the earth temperature is fairly constant. Vapor barriers become more important with earth-buried heat storage, and the insulation selected must withstand pressure (not crush), but the advantages prevail and make it worth checking out.

The maximum insulation needed may be computed from the difference between the worst temperature your area experiences and the highest operating temperature you'll need from the collector to fulfill its purpose effectively. With a backup system, this is pointless and costly; just design for the average low temperature in the worst month and leave it at that. A world of difference exists between these two levels—the worst and the average worst. Designing for the worst temperature is like stuffing a backpack full of everything you'd need to combat every conceivable situation. You'd be well prepared, but wouldn't be able to budge. Cost isn't the only consideration; other factors limit the practical size of an installation. For all-time low temperatures, your efforts are better spent struggling into a sweater.

CODES ON SOLAR

As with any other thing that you add, delete, or improve upon in the dwelling you call home, house, shop, garage, storage shed, or barn, you are supposed to have a building permit for a solar utilization installation. If you're remote enough, or you have neighbors that support your efforts, you may not need one. Of course, if you ask at the building department, they're going to say you do, because that's the way they earn their living. There is a fine for construction without a permit, but the bottom line is "if you get caught." The storage tank is usually no problem, since it's inside, but the collector can't be hidden in the cellar. The price you pay for a permit normally will not be excessive, and certainly worth the money if you're going to lie awake nights, develop an ulcer, or become paranoid. On the other hand, it may spice up your life to take a risk and get away with it. Take your best guess and go for it.

Roof loading (if that's where you put the collectors), plumbing considerations, and storage-tank siting will prove the only possible trouble areas in getting your plans approved inexpensively. An old roof may, in the opinion of the building inspector, be incapable of supporting the collectors. Since most collectors don't weigh more than a few pounds per square foot, they won't be the problem; it's the people installing the system that may exceed the roof loading! And, of course, the inspector may be right, in which case you have two alternatives: beef up the roof, or locate the collectors elsewhere. New roofs should receive immediate approval, unless your collector is made of concrete; it may get by the inspector but you'll probably end up with it as a bed partner some crisp early morning.

Plumbing's no problem. The inspector doesn't want it to leak and you don't want it to leak. That takes care of that. Make sure the inspector understands you have a double heat exchanger for the glycol, if applicable. Don't forget to point out the T&P valve or valves in the plans of your system; inspectors like

44 SOLAR ENERGY

that a lot. Siting the storage system (rock, water, etc.) poses little problem in a new house, because the specific site can be engineered to handle the weight. It's tougher for a finished home, but the examiner will know what the house can or cannot take. If you don't want to take this individual's word for it, have the structure independently engineered or get a professional opinion; this is worthwhile in borderline cases, where the inspector doesn't want to take the responsibility. Otherwise, compromise, locate elsewhere, or beef up the intended site.

Ever hear of the sunshine law? This should be more in the news in the years to come as people employ solar energy in home and business. Simply stated, you have a right to sunshine on your property for a significant portion of the day. If your neighbor wants to build something, he may be given the choice between modifying his intended structure or abandoning it if it's going to interfere with your solar access (in other words, shade you). You don't necessarily lose the case if you don't have collectors up and operating, because light is necessary to health and well-being, too. However, this law, while in effect, may never be enforced, at least at the building-department level—heck of a lot of good it's going to do you to complain after the building's in. For this reason, you'll probably have to sue as soon as you see what's happening, and, of course, your chances of beating even a fairly big corporation are much better if you're actively using the sun's energy. No, not suntanning—using collectors for cooking, heating water, distilling, or whatever.

Collector arrays don't necessarily have to be laid flush against the roofing surface. Bracing can be added at the upper or lower edge of the collectors to extend either of these areas outward if the roof angle is too steep or too shallow compared with the computed angle of declination. Detailing this assembly may provide some adjustment to the angle, in the desired direction at the correct time of year, for optimal use of the sun's energy.

For this situation, and ground installations, a word of caution is in order. It may be skipped by the building inspector or you, if you're building sub rosa, but you don't want to ignore the effects of wind on your assembled collectors. Even relatively light winds will expend a lot of energy trying to get by your cantilevered collectors. If the wind is unsuccessful, all we get is buffeted air. If successful, you've got kindling for your fire or a mess to clean up. Wind is powerful stuff; if you doubt it, check how much energy you can get from it (Chapter Three). Bolting a collector to the roof only compounds the problem—it does not relieve the forces exerted on the exposed surfaces, and you may lose both roof and collector. A low profile is the best bet. Flashing made from metal or wood

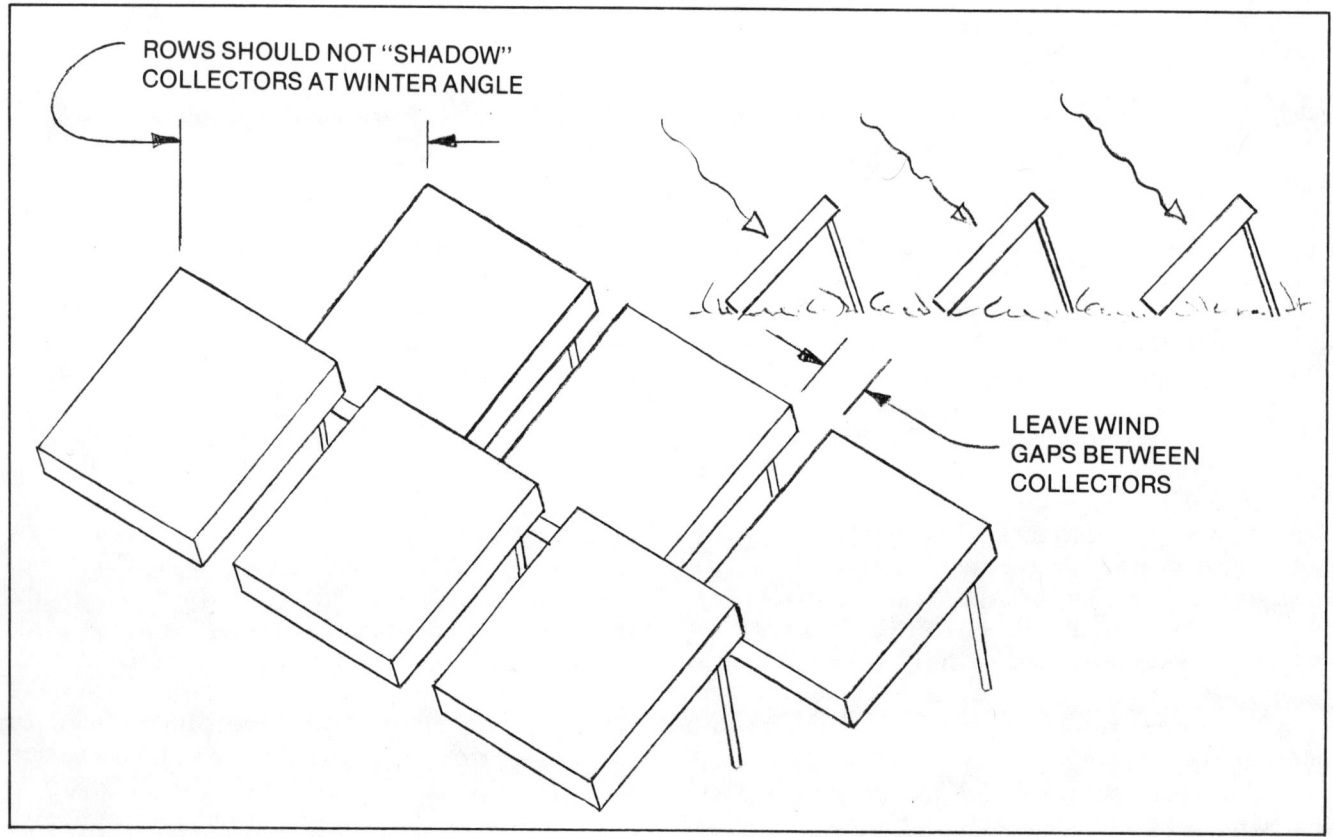

Fig. 2-45. "Sawtoothing" the collector array

can smooth over areas where the wind gets behind, under, or generally trapped. With ground installations, staggering or spacing the collectors will relieve much of the wind pressure if the predominant wind is from the same direction in which the collectors are oriented. If the wind comes from the back, a sawtooth arrangement of collectors (see Fig. 2-45) is the only deterrent to their sailing away in the first big blow.

BACKUP HEATING

I've said this elsewhere, but it's worth repeating. Particularly here, for this is the reason for backup heating—a solar-utilizing system which is designed to deliver 100% of your heating needs will cost at least ten times as much as a system that delivers only 90%. That last 10% is a wallet-sucker because it's the linear component (y-axis) of an exponentially rising curve. For the non-trig readers, that means it goes up real fast. It makes sense to use a backup heating system for this reason alone. There are other reasons, though. What happens if you need to work on the collectors, storage springs a leak, a blower motor burns out? An auxiliary heating system buys you time.

Alternate sources of heat abound. There's prompt propane, easy electricity, and crusty coal. Fuel oil, kerosene, and natural gas are others. Buy and burn chemical logs. How about pocket warmers? Sweaters? A blanket? A close friend to share the blanket? Jogging? Then there's always firewood, too.

Picking the heat source that suits you is an individual matter. Pick a versatile source that's readily available, and useful. Pushups are great for body heat, but they won't get the shower water warm, much less hot. It helps to choose a renewable resource. That way, when the nonrenewable ones are no longer available, you're not affected. If a major amount of your energy comes from a solar heating system, don't feel any guilt about using gasoline, propane, or store-bought electricity as a backup. The other alternatives —besides the one of just doing without—seem limited to methane, windpower, small-scale hydroelectric power, and wood. To all but city dwellers, however, wood offers the next best alternative source of heat to solar. (Check out Chapter Four.)

When it comes time, try to merge systems, avoiding duplicated parts—blowers, ducts, water lines, etc. Some things don't become apparent until you start to build, but the plan should allow for some variance or versatility in the system. But you've got to have the plan. Grab a favorite drink, sit at a comfortable counter, table, or desk, and use up a few pads of graph paper. Try this in different atmospheres and different places until one way works.

SPECIFIC APPLICATIONS

At the beginning of this chapter I teased you with a variety of applications for solar energy, yet we've been dealing mostly with space-heating and water-heating systems. In the following sections, I plan to get more specific about these and other uses. I haven't provided actual plans, but, for the more enterprising, my descriptions will be a nice substitute. And, if you don't fit into the enterprising category, the information will help you understand how it's done, what plans will work for you, and how to modify a set of plans to fit your own needs.

SOLAR COOKING

Solar cooking is exciting. Whether you've got a barbecue or an oven built to use the sun's rays directly, it's pure magic. And, like most magic, mirrors are involved. Only we call them reflectors. And organic material is the absorber. That's the food!

Solar barbecue. The simplest solar cooker is the curved dish (see Fig. 2-47). This may look like a parabolic curve, but it's not. Parabolics and cooking don't mix; that's like using a laser to blow-dry your hair. It just so happens that the 4-foot-wide arc of a circle with a 4-foot radius approximates the curve of a parabola with a focal length of 18 inches (you may have to read that over again!). The main difference is that the strongest focus you can get is about 6 inches across. That diffused a concentration permits us to cook something without burning a hole through it.

The dish is easy to build, using all manner of

Fig. 2-46. Collectors on the roof of a restored New York City tenement are "saw-toothed" to prevent them from shading one another, and to ease wind pressure.

46 SOLAR ENERGY

Fig. 2-47. A homebuilt parabolic solar cooker ignites paper within seconds.

materials. Even cardboard and posterboard. The rigid framework and dish is then covered with adhesive-backed aluminized Mylar or, for the hobo version, rubber-cement-glued aluminum foil. Then get the hot dogs (or broccoli if you're a vegetarian), point the dish at the sun, and have at it. Bright, sunny days work better than cloudy, rainy ones.

Baking bread, cooking a roast, or broiling some rabbit takes longer and is more difficult in the barbecue. Hence, the solar oven. It offers unattended operation (for as long as twenty-minute intervals anyway), simplicity in design and function, and little strain on the budget.

Halacy oven. The Halacy oven (see Fig. 2-48) is a container with a window. Put food in the box, point the glazing at the sun, and the food will get warm; on a hot day, it will get hot and maybe even cook. Add some refinements: foil-line the inside of the box so that the rays which miss the food first time around will catch it on the rebound; tilt the box so that the glazing is perpendicular to the sun's rays, maximizing the number of square feet of solar energy received; take the food out of the metal pan and put it into a glass pan—sure, energy absorbed on the pan will heat the food eventually, but let's get the rays to pass through the glass and strike the food directly. Now, on a hot day, with an adjustment every fifteen minutes, the food will always get hot and cook some of the time.

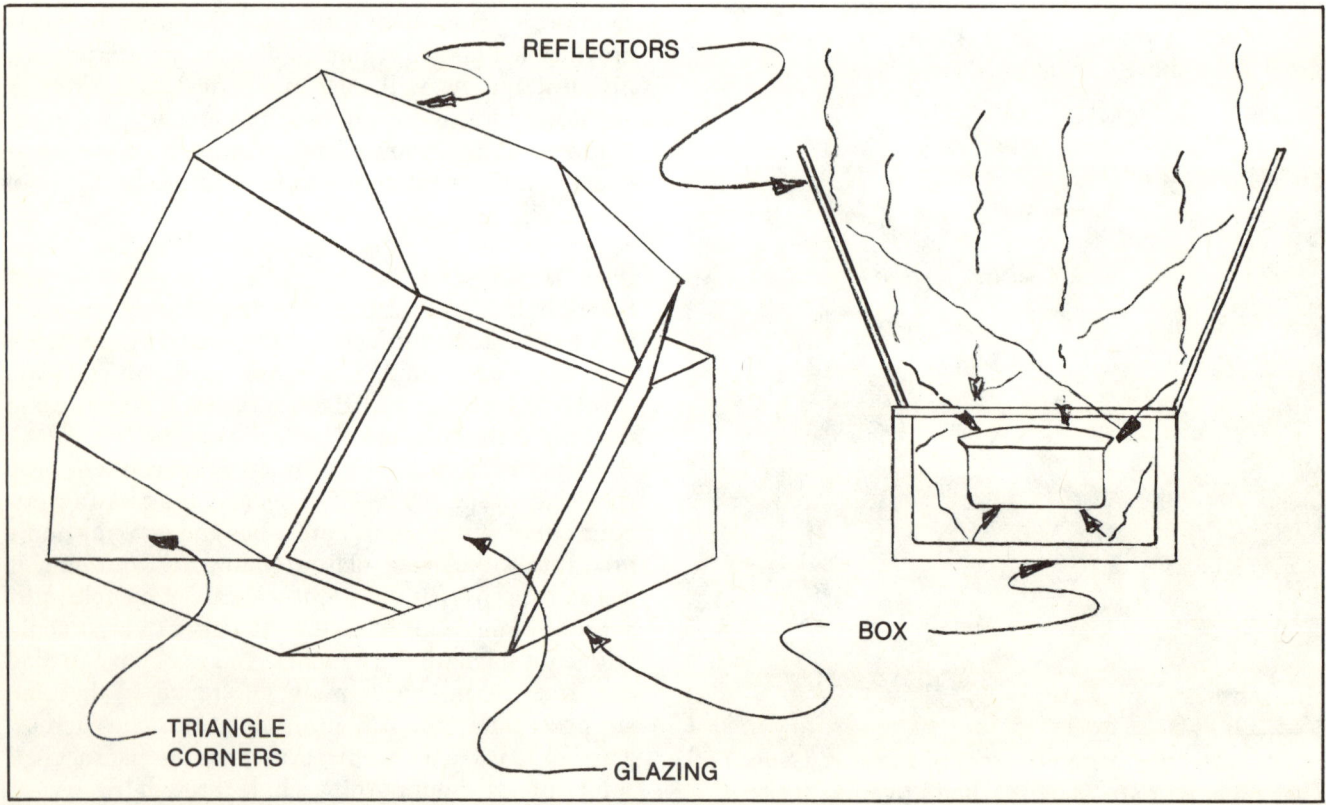

Fig. 2-48. The Halacy solar oven

How can we get some reliability from this unit? If we don't know it's going to cook the food after all our efforts, maybe we should try some other kind of cooker, or another source of energy altogether! First let's try increasing the size of the oven. This should give us more energy. But, alas, we have more materials and more surface area through which heat may escape. Wait! Keep the box the same size, but increase the amount of energy coming into it: use reflectors.

After some experimentation, the best number, size, and arrangement of reflectors may be discovered. To lend some portability to the oven or, at least, a means of closing it when not in use, we do two things: limit the size of the reflector to the size of the glazing, and, with hinges of tape, metal, etc., secure them to the collector so they fold down over the glazing upon one another. To gain the most rays, we find that equal-size reflectors are needed; this dictates a square glazing. When we unfold the reflectors, we find that an angle of 55–60° directs the most light through the glazing and into the box.

There are some gaps around the edges, and we need some way of holding the reflectors in place. One modification solves both problems: fit a triangle of reflector into the gap (on all four corners). Now tape one edge of a triangle to one of the two adjacent reflectors, so that it folds with the reflector. Tape the other edge of each triangle to the other reflectors and the whole affair will assume the correct position. Unfortunately, it's still very ramshackle and, in any kind of wind, may even collapse. To remove the flex, stiff wires (clothes hanger or baling) may be taped at the triangle corners or crisscrossed and affixed to opposing reflectors.

Reflectors added in this manner will always at least double the effective collection area, putting twice as much energy into the box as the glazing would do. The reflectors may be foil-covered cardboard or poster board. Aluminized Mylar on a stiffer surface, like Masonite, works even better. I personally recommend polished aluminum sheet for the serious oven builder (since we know that the bitterness of poor quality remains long after the sweetness of low cost is forgotten).

Access in some of these collectors is through the glazing. Others put it at the back. The trouble with the former is that it's a little awkward, especially if the reflectors are hinged to the glazing instead of the box. The through-the-back oven opening has its limitations, though, if the back has become the bottom due to orientation. A side opening seems to solve the problem. Be certain to turn the collector out of the sun before sticking a hand in the oven, however, with any of these methods; the concentration of sunlight is not too intense, but it could still burn or startle you.

Studies with the Halacy oven reveal that some thermal mass in the oven is more beneficial than none. This will help the oven when the sun is obscured by clouds for a few minutes, preventing a rapid temperature drop. The most suitable material tested was thin brick. If you decide to add this to your design, build bigger to compensate for the occupied space.

Oven comments. With any oven using a concentrated energy, wear sunglasses or welding goggles when working around the front end. Since orientation is important, devise a means of detecting when the collector is zeroed in, other than by sighting (with your eye) along an edge toward the sun. The shadow technique is a good one. Epoxy a golf tee or a 2-inch piece of wood dowel perpendicular to the glazing. If the sun strikes anywhere but straight on, you get a shadow. Simply rotate the oven until the shadow disappears, and the rays must now be striking perpendicular to the glazing. If there is no shadow, the sun's light is probably too dim to do much cooking anyway.

As you'll recall, the sun's position changes at the rate of 1° per four minutes (derived from 15° per hour). The reduction in energy to the oven is noticeable after 4° to 5°, which means a change every twenty minutes. You can stretch this to a half hour if you advance the oven's orientation a little beyond the sun each time you move it, and allow the sun to catch up.

The Halacy oven may be made of cardboard, styrofoam, wood, or metal. The cardboard one is great as a children's project and it's one way, without expending a lot of energy, to get one built. Easy-to-build means easy-to-break in this case, however; leave it out in the rain once and that's it. Two-inch

Fig. 2–49. Ron Alves removes the door of a Halacy solar oven.

Fig. 2-50. A large-capacity solar still
(Courtesy James L. Ruhle & Assoc.)

foam sheets make great but short-lived Halacy ovens. Internally, the foam must be faced with poster board before the foil is applied, or the oven will melt. Even so, after a season in the sun, the styrofoam warps with the heat and chips with the abuse; it's useful for the experimenter, but can't take the whipping that a metal or wood oven will. The wood oven, while not as "eternal" as the metal collector, requires very little insulation, if any, and is much easier for most people to build.

The glazing, irrespective of the type of oven, is glass. The temperature of this oven can easily reach 500°F and will often, during normal operation, hover at about 300°F. Any other type of glazing I've used could not hold up or allow this temperature inside. Ensure a tight seal around glazing and door seals (if they're not combined); with this high a temperature, heat is going to be coming out the pores. If the glazing fogs, it's just condensation of the vaporized water from the food; a few small holes drilled in the sides and back of the oven should help alleviate this.

SOLAR FOOD DRYER

A food dryer liberates you from installing a large freezer and paying to power it; drying food is preserving food, and a smaller refrigerator can take care of the remaining perishables. Drying food in a collector that's designed for that purpose has many advantages over letting the sun beat down on spread-out produce. First, it speeds up the process. Second, it keeps some crops out of the direct rays, which can be harmful. Third, it keeps rain from spoiling your efforts. Fourth, it minimizes the effect of windborne dust and other contaminants. Fifth, it keeps your cats or dogs and sneaky children or adults from making off with the goodies. Sixth, it controls insects, both the flying and the crawling kind.

A flat-plate style collector works wonders as a

Fig. 2-51. A solar food dryer

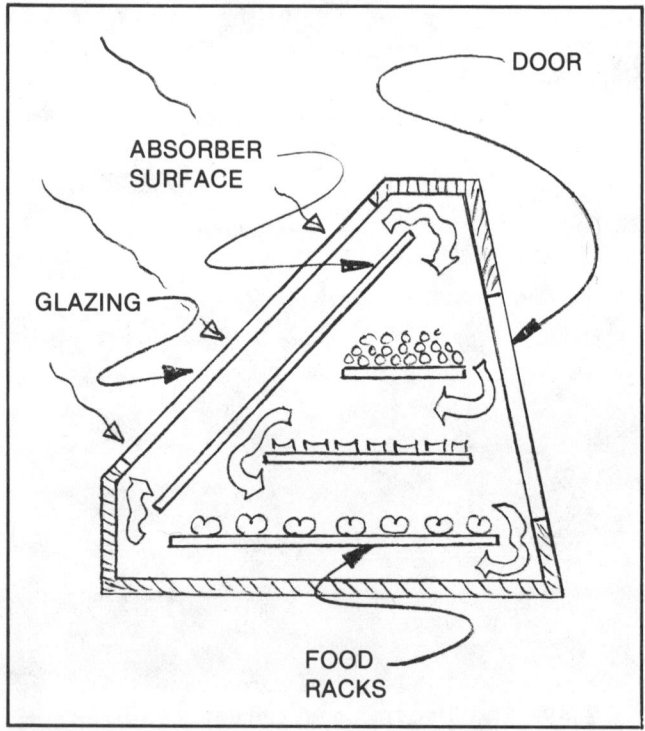

Fig. 2-52. A large-capacity solar crop dryer

food dryer. Make it deep enough to handle the food, which is spread on a grill or screen, insulate it, paint the inside black, and attach the glazing. Oh, and don't forget some way to get the food in and out! The difficulty with this design is that in order to dehydrate the food, we want to remove the water. Unlike cooking, we don't want to vaporize the water—just evaporate it. But where will the water go? If we don't have some way of removing the moisture, we end up with a solar still effect—the evaporated water condenses on the glazing. Then it blocks incoming energy and falls back inside to evaporate again. This cycle will soon shut down the dryer. To alleviate the problem, holes must be drilled in the sides to allow the moisture to escape. Of course, the heat will use these exits as well, but that can't be helped; the changeover of air is necessary lest it become saturated with moisture. Insulation is still used extensively, minimizing the heat loss.

A design variation that allows a larger capacity is a modified dual-action collector (see Fig. 2-52). In this design, none of the food is directly exposed to the sunlight, because it's mounted behind the absorber sheet. Instead, air is heated and circulates around the food, which is spread out on screen sheets (plastic screen won't rust) in racks like dresser drawers; if a metal absorber sheet is used, some of the food will be heated by reradiation from the back side. A small vent shutter at the foot of the collector can be rotated to circulate the air inside the collector or to take in new air. If outside air is being heated for use, the vent holes in the body must be large enough to vacate the moisture-laden air. Put some fine-mesh screen over any holes to keep the bugs out (the aroma of dehydrating food will bring 'em from miles away) and use bug-tight construction throughout. Although many insulation materials can work, fiberglass insulation is recommended, since it won't support insect life. Treat all wood and don't set it directly on the ground. If you use stain on the inside, let it air in the sun for some time before using—there's enough bad stuff in our food already.

Temperatures above 200°F are unwanted, so adjust the airflow to keep it below this level. In cooler climates, double glazing may be required; allow some air circulation between the glazings, or condensation will spoil the efficiency. Glass is recommended except where children play—that will require fiberglass or plastic glazings Be sure the glazing can take the UV, the temperature, and the exposure to moisture.

GREENHOUSES

There's one thing better than flying a sailplane through thermals high in the sky, and that's walking out into

Fig. 2-53. The DeKornes' "survival greenhouse" and hydroponic jungle. This pit grow-hole makes use of the good insulative properties of the surrounding earth.

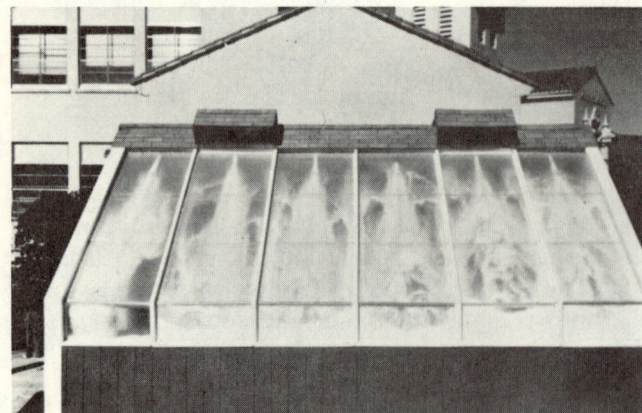

Fig. 2-54. Plastic insulative beads are being blown between the double glass on a Beadwall greenhouse designed by Steve Baer.

the attached greenhouse and having a morning cup of coffee. The glass to use is translucent (not really transparent), so there's just a hint of sun, sky, clouds and trees, but it's all green inside except where there's a blob of yellow or red, or a string of orange or mottled green, and a sprinkle of purple and blue that eludes the eyes' attempt to focus and define. Besides the heady effect of all that oxygen, and the wonderful assault on the nose, and the urge to throw off all your clothes, there are a few practical reasons why a greenhouse should be part of your life.

Severe-weather areas need greenhouses for year-round food production. The most widely known application of greenhouses today is for growing flowers throughout the year, so there are always flowers to soothe that midwinter spat or sparkle the eyes of your prom date.

The term "greenhouse effect" comes from using greenhouses (of course), and the principle applies to any collector that uses glass—incoming sunlight is passed but outgoing thermal energy is stopped. The net effect is a warming of the interior during exposure to sun. At night, however, the effect is canceled, and it can be mighty brisk in there come morning. Therefore, greenhouses need help at night from other heat sources. Or some of the incoming energy in the day can be tucked away for use at night; water-drum storage is one of the better techniques.

Even heat storage finds the going tough on cold nights. It's all that darn glazing, and the black sky is a real heat sink. So, cover up the glazing. Covers may be shades that are drawn, styrofoam panels hinged down and fastened, or a Keppernau opaquing-field screen. Or do it the way Zomeworks did it with the "Beadwall": blow tiny polystyrene plastic beads between the double glazing. The whiteness reflects radiated energy, and the air/plastic combination gives an R-value similar to fiberglass insulation. When you

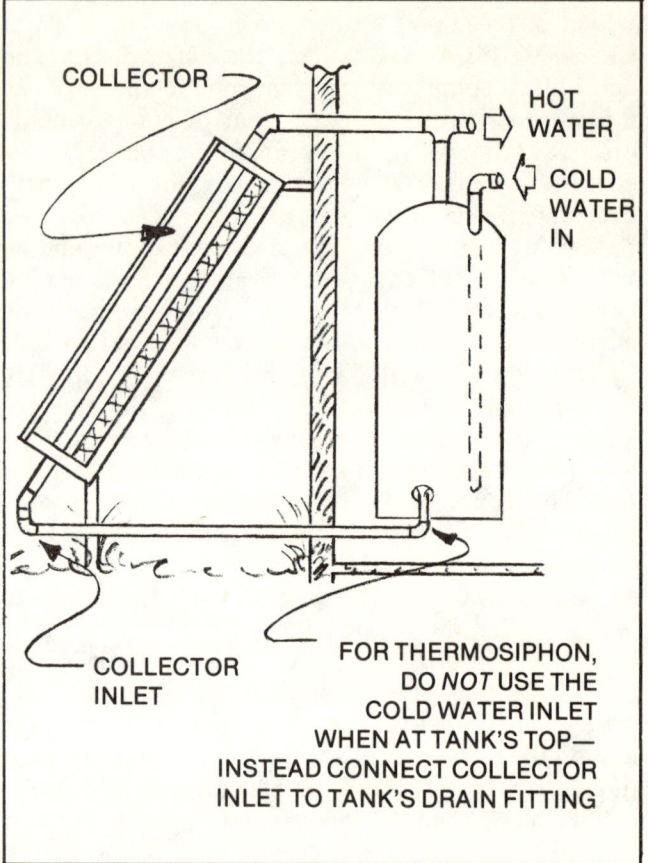

Fig. 2-55. Adding a solar collector to an existing hot water system

want the sunlight in, you suck the beads back out using blowers and store them in drums.

The well-designed greenhouse uses combinations of these techniques to assure adequate heating for all but the most extreme situations. Glazing is minimized to areas of maximum benefit, and nonglazed portions are well insulated. Thermal storage and sliding, swinging, or hand-insertable sections are used over the glazing, as needed. Layout and architectural angles become just as important at keeping out heat as retaining it. Vents should be installed to aid in getting rid of accumulated heat in the summer. Even daytime temperatures in the winter can get too high; the attached greenhouse, then, may help heat the house. If the greenhouse size is exceptionally large, it may even be designed as the collector for winter space heating. A backup heating system is recommended for peak load needs.

To keep this habitat from becoming a steaming jungle requires patience in design and construction. Whole books have been written on the subject (see the Sources and References section), and this information should be studied before designing your own greenhouse.

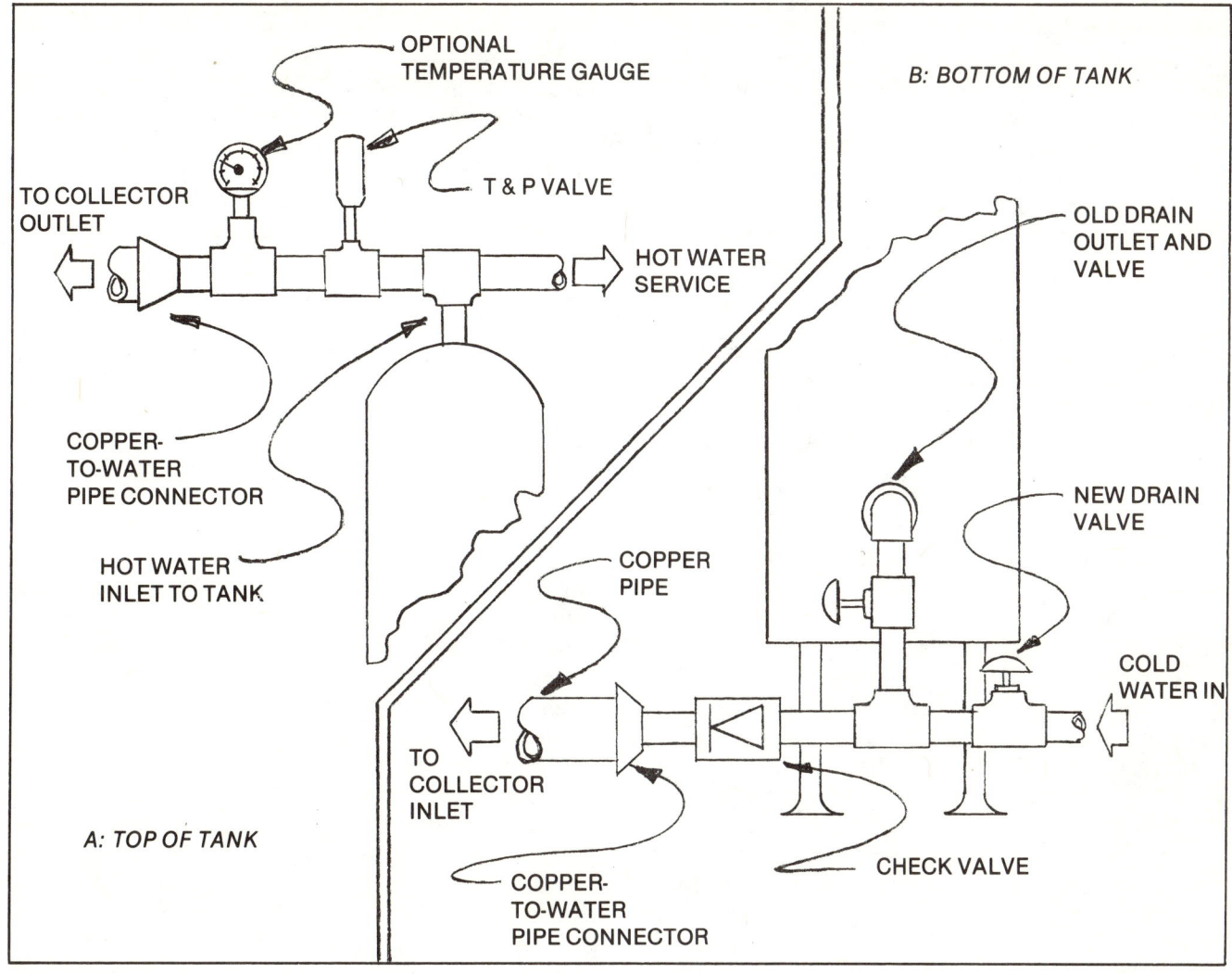

Fig. 2-56. Collector connections at the water tank

WATER HEATER

You've decided to solar-heat part, or all, of your water. What now? First, let's put together a simple design (Fig. 2-55). You've got a 50-gallon hot-water tank that's located with an outside access on the south side of the building; let's use that for your solar storage. Since the collector will be located at ground level until you've passed the "experimental" stage, we'll use a primary system and natural convection (thermosiphoning) to move the water from collector to tank and back. That means we'll need a check valve in the connecting pipes too, to prevent reverse flow at night. Build your collector to the specifications previously noted in this chapter, including information from other sources and your own preferences.

At the hot-water tank, locate the inlet and outlet pipes; the hot-water pipe is usually insulated, but not always (after all, the person who installed it doesn't have to pay the bills). Also locate the drain valve. This will be low on the tank and is the best attachment point for the tube running to the inlet valve on the collector. Put a T-fitting there (Fig. 2-56A). Put in a short nipple and attach still another T-fitting for the collector's T&P valve; if the unit comes with one, check with the local building department to make sure it's adequate. You'll need some copper pipe to connect the tank with the collector. Once you've set your collector, measure this distance, get the pipe, and cut it to the right length. Get a reducer that will mate the 2-inch copper pipe to whatever is used at the tank; this will be the bottleneck in the system, but better here than elsewhere. After you've done both top and bottom, finishing any required soldering, turn off the cold water inlet (at the main or right before the tank, if there's a valve). Disconnect the hot water fittings, insert the T-fittings and the T&P valve, and attach the outlet pipe of the collector to the hot water (outlet) pipe of the water heater. Connect the collector inlet pipe to the drain valve on the tank (Fig. 2-56B), open the old drain valve (make sure the new drain valve is closed), and check the fittings. Go open a hot water valve somewhere in the house and then open the main or cold water inlet valve into the water heater slowly. This should let water fill the collector and evacuate all air bubbles. Once the sputtering's stopped, you're in business. If the collector is getting hot and so is the outlet pipe at a point several feet away, you've got thermosiphon flow.

The interconnecting pipe will need insulation and waterproofing. The old drain valve allows you to shut off the flow of water through the collector, but a check valve should be installed here for automatic protection against reverse flow at night. *Never* shut off collector flow without providing some means of preventing the apparatus from collecting sunshine; have a cover handy for the glazing if you need to do some work.

Somehow the present water-heating source—electricity or gas—must be disabled to give its competitor a shot at heating the water. If it's electric, you can pull out the breaker in the morning when you leave and throw it back on when you get home. A gas heater will have you on your hands and knees, turning the thermostat from "main burner" to "pilot only." Better yet, rotate the knob all the way to "off"; even the pilot generates a lot of heat, wasting gas in the process.

It'd be nice to know how hot the water in the tank is when you're ready to use it in the evening; a temperature gauge mounted in the pipeware would be a help. Because water of varying temperatures will stratify in the tank, locating it at the top might give too high a reading and one at the bottom too low; the middle point would be best but, alas, there's seldom any kind of opening there. If you're an evening-shower person, have at it; if it's not hot enough, throw in the breaker (or put the thermostat back to "main burner") and let it top off the temperature. Various devices can be installed to automatically kick in or kick out the tank's own heat source to ensure the right temperature of water when you want it, but you'll quickly learn (through use) what works and what doesn't, and save yourself a lot of money without them.

Once the system is proved and no leaks are detected, insulate the main tank. There's insulation under that metal skin, but it's very inadequate. Wrap R-11 (4 inches of fiberglass) around the tank and secure it so that it doesn't get compressed. If you want it to look good, use paneling or some other material on top of it. Allow access for the controls and arrange a hose so that water coming out the T&P valve, should it go off, won't flood the place or dampen the insulation.

COOLING TECHNIQUES

Sure, no one likes to be cold in the wintertime. But what about the ravages of heat in the summer? Is there any relief from that? How can we get solar to do a little less in our lives instead of more? I said something at the beginning of this chapter about solar refrigerators and solar air conditioners. Well, these devices do exist. Before you can understand how they function, however, you'll need to know a few basic refrigeration principles. And, since this is far more involved than other cooling methods, you should be convinced, first of all, that the other systems are unsatisfactory for you in some way.

We'll start with the simple and proceed to the more complex. I've got exposure, convective currents, shading, evaporative cooling, night-sky radiation, and the heat pump on my list.

EXPOSURE

Exposure is absurdly simple. If we have a hot something we want to cool, we just place it where it's cool and let the heat-seeks-cold thing happen. If it's too big to move about, we wait until it's cool enough (nighttime?) and then expose it to the environment. In the days before refrigerators, people would take perishable items and just throw them out into a snowbank for the winter—provided wild animals couldn't tear into them and you didn't forget where they were when you needed them, that worked fine. I've often thought it a pity to use up electricity for a refrigerator when it's colder outside the house than inside the icebox; suppose one could just open it up to the outside for as long as the cold snap hung around. There are ways to do this. If we're dealing with a refrigerator, it would work more efficiently if we took the compressor/motor heat coils and put them outside; in both winter and summer, it would be on a lot less since it could better dissipate the heat it removed from the inside. If the refrigerator were outside on the porch, its door could be left open during the winter cold spells or on cold nights at other times of the year. Exposure, on the simplest scale, means that you don't put hot things into the refrigerator; let foods or liquids cool at least to room temperature before entry. Because of its insulation, the icebox cannot take advantage of exposure as a cooling method; once inside, the food's heat can only be removed by refrigeration techniques.

CONVECTIVE CURRENTS

Convective currents does for space heating what exposure does for refrigeration. The idea is simple but, unfortunately, the techniques are not. That's not to say you can't get a cool breeze to go through your house; opening windows can get you that. However, cooling at night involves more than just getting rid of the previous day's heat; it also means storing enough cold to get us through the next day. So, most designs work, but not as well as we'd like them to. This takes some study of what works and what doesn't, and what does what best. I'm not trying to be vague. Take, for example, windows. They sit in the wall, an equal distance from the ceiling and floor (so we can see out sitting or standing), and they open to admit natural breezes or vent heat. Only they're too low to get rid of the heat that's near the ceiling and too high to circulate air near the floor. Moreover, much time and money must go toward the perfect seal lest cracks around the window leak drafts and allow heat to escape in winter.

I'm reminded of a story I once heard. I can't specifically remember how it went or where it came from, and I've added a new twist, but I think it will give you the feeling it gave me when I first heard it.

Once, a man lived in a bamboo house. The bamboo stems were tightly woven, but they let in just enough light and just enough air. In spite of that, an improvement seemed in order to stop the air that moved into the house. So a wall was built. But it shut out the light. So a hole was cut in the wall. Now the air came in. So glass was cut to fit the hole and installed. But when the sun shone in, the house became hot and stuffy. So the glass was hinged to open. However, the light was strong, so curtains were added. Alas, this blocked the flow of air. Our man bought and inserted an air conditioner. But now there was little light, since the unit took up so much of the window. So electric lights were installed. Now it was cool and light. Suddenly, there was a blackout....

Other than architectural designs, better cooling is often accomplished if vents to the outside are located near the ceiling and floor; with enough of these on each side of the house, they may be opened or kept closed to take advantage of the specific breeze that's blowing. This means that windows can be nonopening and, even if doublepaned, only a fraction of the cost of openable ones. Plus no cracks to leak air or heat. Blocks of insulation (styrofoam?) can be inserted into the vents come winter to prevent heat losses through these orifices.

SHADING

Hearing that it's 105° in the shade is a common expression that's supposed to make you wonder how hot it is directly under the sun. Conversely, it should signify the value of good shade. As much attention in solar design is needed to limit the effect of solar influx as to maximize it for usage. An awning, then, is a solar collector, intercepting energy that would otherwise impact where it's unwelcome. Or it might be a solar reflector, ricocheting the unwanted energy elsewhere (the neighbor's house?). Gotta be careful, though. Too much shade and maybe you won't get much benefit from the sun in winter. The trick, you see, is to keep out the sun in summer and let it in during the winter. Knowing the angle of declination for your latitude (winter and summer) is the first step in the computation of roof angles and overhangs to block or admit direct sun in the home you are building or modifying.

EVAPORATIVE COOLING

With long lines at the gas pumps, you may have had an occasion to use an Oklahoma credit card (a siphon hose) and, in the process, spilled a little dinosaur milk (gasoline) on your bare hand or arm. Feel that chill? That's evaporative cooling—the same principle as putting a water bag outside your car as you whirlwind through the desert. A little water weeps through the heavy cloth and evaporates, cooling the rest. Or, if you've used a swamp cooler, you notice that water and a blower are mixed to somehow expel cool air. Ever wonder just how that works?

Water, if you'll recall from our previous discussion, needs 1 Btu per pound to hike its temperature up just 1°F. A pound of water at 60°F would demand 152 Btu's to bring it to 212°F. Subtract the temperature differences and that's how you get the number 152 for an answer. Now we're involved in a change-state situation—the water (at standard pressure) wants to become a vapor. Pumping another Btu in there won't get us an increase in temperature—just a teeny bit of water converted into vapor. In fact, we'd have to inject a beefy 956 Btu's into this pound of water before we'd notice any increase in the temperature (to 213°F) of what's left, which isn't water—that's all been changed into vapor (steam). If you forgot to put a lid on the pan, it's all gone.

We don't have to boil water to get vapor; if that were a necessary process, it would never rain (the ocean doesn't boil). Left alone, water will liberate itself into vapor. Given the correct circumstances—very dry and fast-moving air and higher temperatures—water will vaporize at a rapid rate. The process still goes on even if it's humid and cooler, but much more slowly. Nonetheless, when the water evaporates, it takes heat to do it, and grabs it from wherever it can. If there's just a lot of water around, the surrounding water donates its heat and, in the sacrifice, gets cooler. If there's fast-moving air, the air winds up cooled. Drop this through ducting into the house and stand in front of the vent and you, too, will be cool.

But (you knew this was coming, didn't you?) there's a price. TANSTAAFL! Air cooled by evaporation is laden with moisture, and it feels cold and clammy. All the same, wringing cold is welcome relief to dry hot in the sun-baked minds of desert folk. YCHE, folks (short for You Can't Have Everything).

On a larger scale, water may be sprinkled on a hot surface such as a roof, to cool it (and the house underneath). With a hot roof, much of that water is going to be vaporized and more will quickly evaporate. So we don't just get absorbs-its-weight-in-excess-heat relief. This is a real suck-up-those-Btu's action. We just put a pipe with regularly spaced holes up on the highest point and with a tiny pump down on the ground, create a half-fog, quarter-mist, and part-dribble for rooftop cooling. That dispenses with dust which dulls the reflectivity of the roof (if it's twin-rib aluminum) and minimizes nail-popping thermal expansion and contraction. Doesn't this sound suspiciously like the drip system described in the section on absorbers? Well, yes and no. This works that way but, with the higher temperatures involved, we don't get heated water. Little, if any, of the water makes it to the rain gutter. But we do get a cool roof and a happy house.

Is it worth the effort? If you can afford the water, yes has to be the answer. Less than 2 gallons of water, used correctly, in a swamp-cooler setup (fast-moving air and saturated-water screens) have the capacity of a ton of refrigeration; that's 228,000 Btu's of heat removed per day. I haven't computed the effectiveness of the roof bath (shower?), but it's certainly in the same ballpark.

NIGHT-SKY RADIATION

Has a science-fiction ring to it, doesn't it? Night-sky radiation. Actually, the earth is always radiating energy; it just doesn't become noteworthy until we have all the right ingredients—a clear, cool, cloudless night with low-humidity air. Stick a blackened pan with an inch of water in it out under the stars, and get to it before the morning sun and—whatdayaknow!—a block of ice. And it only got down to 40°F last night, too! This is night-sky radiation. That poor old dumb heat tries to share itself with the universe (the black sky). Even when it's reached ambient temperature (with the surrounding air), it doesn't see any heat coming back from the big black hole of sky and keeps pumping it away.

To protect most collectors from night-sky radiation and its effect, reverse flow (which loses the heat gained in the day), a check valve is used. The collector, however, will keep dumping heat away and get progressively colder for its effort. And, even though it's only 38°F, the water in the collector freezes and bursts a pipe, much to your surprise and general unhappiness. Heat-collecting systems can, with proper design, use their collectors as heat-dissipating devices by bypassing or removing the check valve. If glass is used as a glazing, the effect is less pronounced because of the opaqueness of the glass to radiated thermal energy. If, on the other hand, you couldn't afford the high-priced spread and glazed instead with Mylar, cheer up; Mylar has a high transmittance of normal heat.

Night-sky radiation can be used to make blocks of ice which are retrieved in the morning (before milking the goats) and deposited in the proverbial "icebox." It won't always work, but it works well enough for the expended effort when the conditions are ideal. Living in a mountainous area that experiences the conditions quite frequently might make a special refrigerator—an insulated box with a water reservoir on top—worth constructing (Fig. 2–57). In the morning, instead of removing the ice, you'd just cover the water/ice chamber with an insulated lid and wait until the next evening.

A hermit friend of mine residing on a high mountain in California told me (via amateur radio) of the best performance results he's gotten using night-sky radiation cooling. His 4-by-8-foot wooden pan with 1 inch of water, on a cool evening (40°F), clear sky

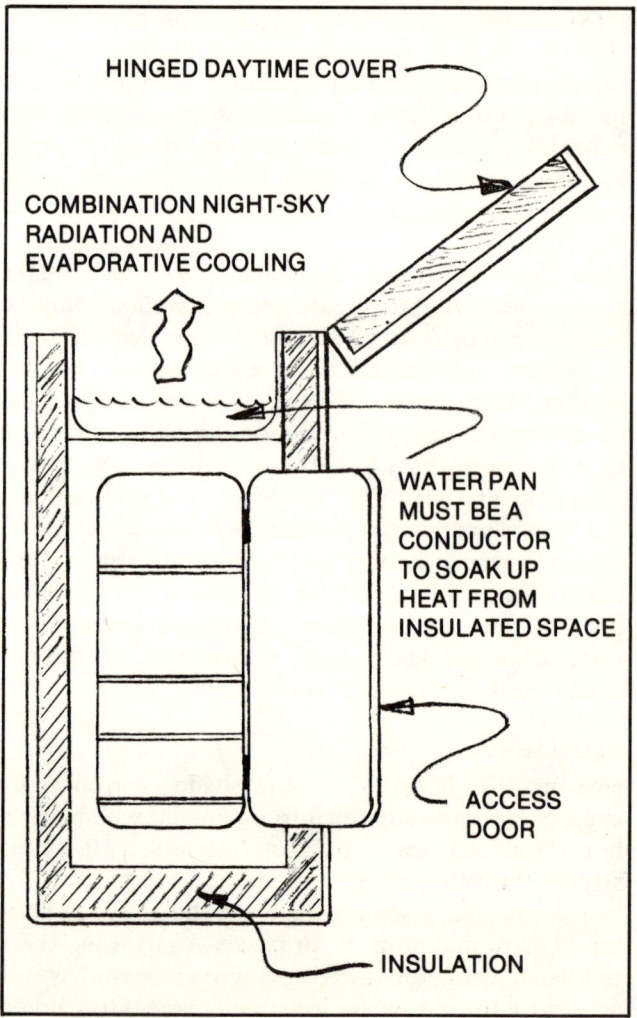

Fig. 2–57. A night-sky refrigerator

and 60% humidity, had a ¼-inch of ice in it the next morning which, broken up and installed in his icebox, came to a total of 40 pounds of ice. He also reported getting ice even when the ambient nighttime temperatures were as high as 55°F, although he only got 6 pounds of it. Check the Sources and References Section for articles on similar units and their reported effectiveness.

Clearly, with the still-low price of propane or electricity, few people will pursue the work involved, however minor, to chill the brew, milk, or ice cream this way. Be that as it may, it's an exciting concept which might prove fun to experiment with. Besides, when we run out of dinosaurs, this method will be in.

HEAT PUMP

In nature, heat seeks cold—which also says that heat seeks places of lesser heat. Leave it to humans to figure out how to get heat to seek more heat. That place gets hotter but the place it's coming from gets cold. Not for long, of course; heat will see this place of lesser heat and rush in. If we insulate the cooled place so that other heat doesn't immediately fill the void, we have a place that gets cold and stays cold, and we could call it, among other things, a refrigerator. Use a refrigerant and the same space can get much colder yet; we'd call this a freezer. Insulate a room and use the heat pump and we'd have a walk-in refrigerator/freezer. Insulate a house and use a still-bigger heat pump and we'd have air conditioning.

Most of us have some experience with refrigerators, knowing they plug into the wall. So, our heat pump must work from electricity, right? Yes, some do. RV (recreational vehicle) owners and campers know they have to light a propane burner to get their little refrigerators to work, so those work off heat, right? True again. Old-timers and the better-informed know that kerosene and natural gas refrigerators preceded the electric refrigerators of today, and the source of energy was an unusually small (it would seem) flame for such a large amount of cold in the inside. Considering that we can achieve some high temperatures with solar energy, if the old Servel crowd and the solar crowd ever got together, there'd be a marriage of the old and new—solar refrigerators!

Understanding how this works isn't complex, but it seems a little magical at times and takes a little study. Unfortunately, refrigeration and heat-load theory is beyond the scope of this book (see the Sources and References section for some good material on this subject). However, I can't walk away from this altogether and leave you in the cold, so here are some tasty tidbits to steer you in the right direction.

Refrigeration principles. I'm sure you've heard the term STP before. No, this isn't what you add to your car's gas or oil. STP stands for Standard Temperature and Pressure—atmospheric pressure, that is. We go through most of our lives at one pressure and know that water boils at 212°F and that's that. Or is it? What happens when you cook that roast in a pressure cooker, beside the fact that it takes less time and is cooked more evenly? Believe it or not, you've raised the boiling point of water by increasing the pressure in its environment. And another thing—remember hearing how the climbers of Everest tried boiling their water for tea but it was only lukewarm? Can you imagine water boiling away but being only just warm? You could if you lived in a low pressure (thin air) environment because it lowers the boiling point of water and other liquids.

Our world is chuck full of things which change state. Here, however, we're only concerned with gas-to-liquid and liquid-to-gas changes. As we've already discovered, when things change state, weird things happen. A substance which required only one Btu per pound to increase its temperature now needs a phenomenal amount of Btu's to go the same distance at a "change state" point. The crucial distinction is that when liquids become gases, large quantities of heat are absorbed, and when gases become liquids, heat in large quantities is released.

Understanding that many different substances have boiling points far above or below that of water, we're ready to imagine a hypothetical refrigerator. First, we pick a liquid with a boiling point (at one atmosphere) *below* the lowest temperature we wish to achieve in the refrigerator. As difficult as it may be to conjure up the image of something boiling at 20° or 30°F, it's only because the mind tends to associate boiling with the term hot. If we pick a substance that boils in the range of 5–40°F and possesses a number of specific characteristics, it's called a refrigerant.

When we've picked the refrigerant for our paper refrigerator (see Fig. 2–58), we then reduce pressure in its environment until its boiling point is much lower. Thus, it boils (or vaporizes), absorbing heat as it changes state (from a liquid to a gas). Insofar as it's happening inside the icebox, it absorbs heat that's trapped in the insulated interior (you just put in a big pot of stew that's cooled to room temperature). Move the refrigerant outside and pressurize it (with a pump) and its boiling point goes higher than the temperature it's at, so it condenses, giving up heat (to be dissipated in the heat coils at the back of the refrigerator). Now, reduce the pressure (the job of the expansion valve) and the cycle repeats. Simple, huh?

If this is the first time you've been told how the trick works, you're not going to get it with the first reading. Read the previous four paragraphs once again. Take it slow, point by point, and the conclusion is definite. Of course, this tells you *what* is done

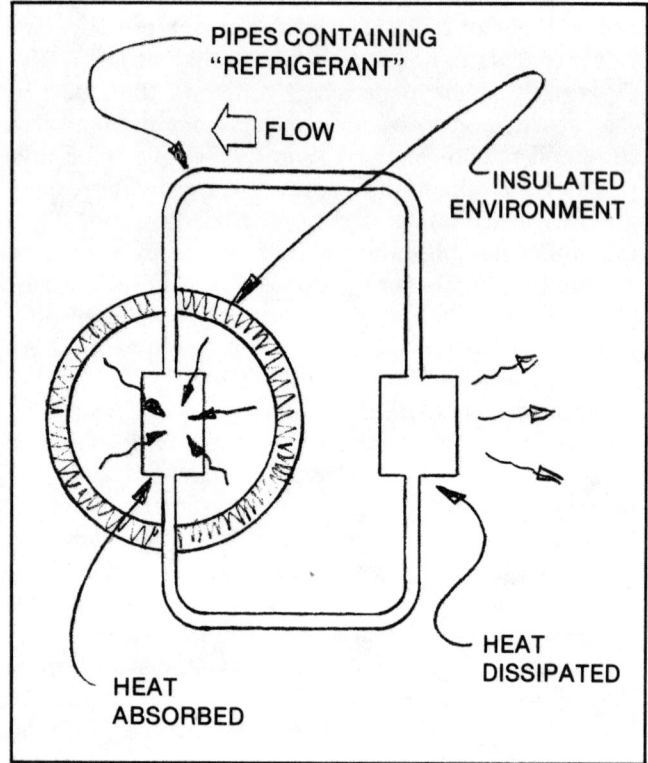

Fig. 2-58. Refrigeration principles

and not *how*. But we're coming to that....

A Simple Refrigerator. It's only so much theory until we involve some hardware and build ourselves a real refrigerator. The closest thing to a solar refrigerator that remains simple enough to discuss at this level is one which resembles the "icy-ball" refrigerator used in the early 1900s. While other heat sources—coal, wood, oil, and gas—were commonly used, this modest unit would lend itself nicely to the temperatures reached in a dish concentrator (4 feet in diameter) with an 18-inch focus. For simplicity of the explanation, I'll assume that we use a small wood fire. In every other respect, the sequence of events is identical with the operating procedures for the icy ball. (Caution: Do *not* try to build one of these units just on the information given here, as a few small but essential details are omitted to reduce the technicality of the description—see the Sources and References section for further info.)

To build this unit we would take two 10-inch hollow metal spheres and join them with a tube (Fig. 2-59) so that a liquid could run from one side to the other freely. Since we're planning to pick up this device and move it around and we don't want to become confused as to which side is which, we'll label one of the balls Y and the other Z. Now, we fill the Z side halfway with a solution of ammonia (playing the role of the refrigerant) and water. We tip the balls so that all of this solution runs into the Y side.

Grab a bucket of water (preferably as cool as it can be) and, setting it beside the little wood fire, put the Y ball over the flame and submerse the Z ball in the water (Fig. 2-59A).

Ammonia boils before water does, so it becomes a vapor which travels from ball Y to ball Z, where, coming in contact with the cooled ball, it condenses. This goes on for a few hours until all of the ammonia is driven from the solution, leaving hot water in Y and condensed ammonia in Z. At the end of that time, we pick up the apparatus and plop the Y ball down into the water (Fig. 2-59B), where it quickly cools and drops in pressure. The lower pressure reduces the boiling point of ammonia in the Z ball and it begins to evaporate. We wait 10–15 minutes for it to really get started and then we stick the Z ball into a closed environment (like an ice chest) which allows the interconnecting tube and ball Y to stick out (Fig. 2-59C). As the ammonia evaporates, it absorbs heat and cools the interior of the ice chest. To keep ice from forming (and preventing good heat flow), the old icy balls has the Z ball immersed in a pan of antifreeze resting inside the ice chest, and this levels off the flow of heat from the interior to keep it from blowing its cool all at once.

The 10-inch sphere size of the icy ball is able to absorb 2,300 Btu's with one charge cycle and discharge (refrigeration) cycle. Pressure in the balls goes to 250 psi during the charging phase as the "generator" (Y ball) is heated. The "condenser" chamber (Z ball) would keep the interior of a 6-cubic foot icebox at 50°–55°F when ambient temperatures outside the box are in the 80°–90°F range. The main disadvantages of the unit are the potentially dangerous pressures achieved and the toxicity of the ammonia itself, which prevents a simple pressure-relief valve from being installed. A case of blowup or get gassed if you leave it over the heat too long. Other than the limited cooling capacity and the time it takes to go through the cycle, it is handy and it works in a crunch, lots better than nothing.

FINAL SOLAR CHECKLIST

This is a checklist for decisions that must be made if you want to install a solar utilizing system. It's a recap of some things already discussed with the addition of a few other items to help keep you out of trouble. It's as complete as my thought processes allow it to be, but you should add whatever additional info applies for you. Because of the interrelatedness of everything on the list, it is not organized by priority. Which means, don't do the first thing on the list and then come back to read the second thing, etc. You're going to have to go up and down the list a number of times when trying to apply it.

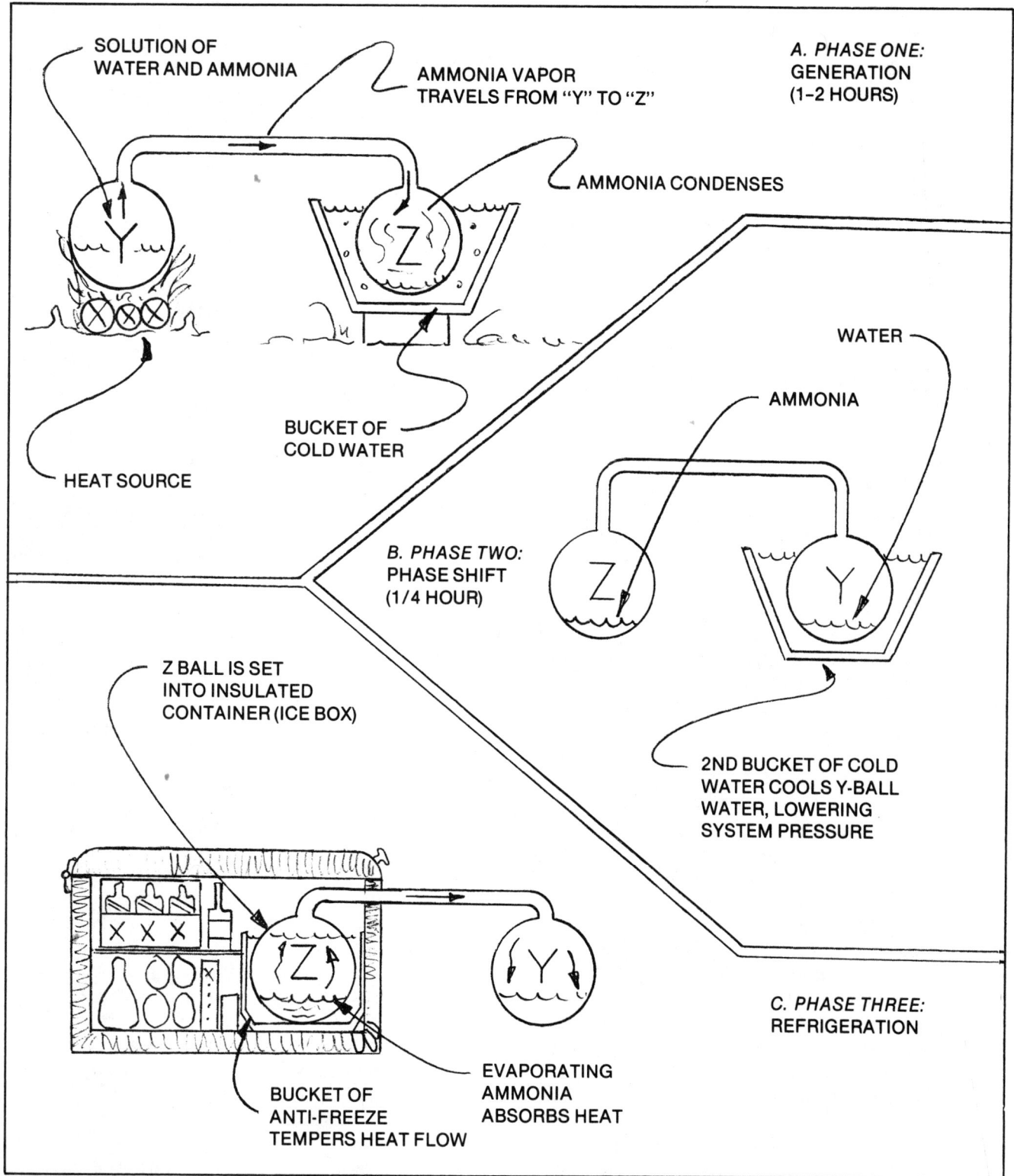

Fig. 2-59. The Icy-ball refrigerator

1. Use. What's the solar energy going to do for you?

2. Type. What collector class (absorber vs. concentrator) and specific type of collector (parabolic tray, tube-on-sheet, etc.) is best suited for the job? Keep it simple.

3. Medium. Air, water, glycol, oil, lighter fluid, carrot juice, etc.

4. Kind of transfer (of the medium). A primary (direct use) or secondary (heat-exchange) system? How does it get from place to place? Thermosiphon or pump? Blowers or fans?

5. Quality of heat required. How high a temperature does the job require? How hot will storage be? How hot will the collector get in normal use?

6. Quantity of heat required. This will affect the size of your collector, in square feet. Or, if you're building add-on units of a given size (4 x 8 feet, for instance), how many will you need? If it's water, one square foot of collector per gallon of water heated is a good rule of thumb. For air systems there is no easy formula; check the process described in Section A of the Data Cubbyhole for computing collector/storage size.

7. Siting. Where do you plan to put the collector array? On the roof or near the ground? Or both? Most of us won't have problems fitting the computed size to what's available, but some may. It's a good idea to know how much space you do have first, so you'll know you have a problem when you've computed total area figures; this saves some wasted computations or design work that stems from the assumption it will fit. Check! This may mean a larger backup heating system, relocating part of the system elsewhere, a different type of collector, moving to another, bigger house, or the final push for you to build your own home. Keep in mind that you must be able to get at the collectors for a periodic dust wipedown or washing of the glazing, or other work and inspection. Most collectors are not strong enough to walk on, so leave roof space above and below for this purpose.

8. Building codes. If you're going to get a building permit, have the system drawn up as a plan (if you do it neatly, this is often sufficient—an architect's drawing is not always required), and present it. The inspector will usually know right off if the thing must be engineered; if the house isn't too old or the collector too heavy (exceeding 6 pounds per square foot), there shouldn't be a problem. The storage tank, if located high in a house, will usually be the problem. Don't forget to show the T&P valve in your drawings.

9. Environmental considerations. I'm not sure what to include in this, but remember that you have to live with this system and so do your neighbors, if you have any. There are many tricks to making it look nice, and all worth the time they take. Whether you want to be or not, you're representing solar energy to the world, unless you were the last on your block to install it. Aesthetics is only part of it. Without getting into politics, each of us has an ultimate responsibility to use less and abuse less. Use low-tech or soft-tech materials and processes. Consider the reflected glare of the collector glazing and whomever it's going to blind for a few hours each day for a few months, etc.

10. Materials. While you're still in the design process, consider the nature of the materials you will use in all aspects of the solar utilization system. Accessibility and availability are very important in material selection. If possible, choose indigenous materials (natural ones found in your area); this saves money, supports local industry, and helps you blend in. The materials should be replaceable; availability fluctuates. An aluminum collector box might last longer than one made of wood, but the latter is a more ecological material and, should you later decide to add on, wood is more apt to be had at a reasonable cost than aluminum. Parts of the system wear out or get broken; a specialized material or function is less replaceable than one that's not. Above all, get material that *you* can work with.

11. Skills. Can you do what you've set before yourself? Skills, experience, and knowledge are important to the successful completion of any project. Owner-builders, beware! What kind of time can you devote? Do you have the tools, or access to them? Do you know how to use them correctly? Do you have the specific knowledge required or know where to get it if you don't?

12. Service life. You should try to figure the earliest date you'll have to replace the system or its parts. The years have a way of slipping by, and when things start to go, you won't be thinking you've run into bad luck if you realize how long it's been since you installed it.

13. Safety. Hopefully you've incorporated lots of safety devices in your system, but this category is more inclusive. Is there an exposed part where an inquisitive child can get burned? Can a streaker run into the collector glazing? Will you fall off the roof when dustmopping the collector? If the T&P valve pops, where's the steam and hot water going to go?

14. Controls. Most systems, particularly those using thermosiphoning, have little need of controls beyond a T&P valve. Winter settings of the correct tilt will prove sufficient for temperature control in the summer months by mere reflectance losses. Other types of shading—roof overhangs, blinds, etc.—will aid, too. In other other types of systems, pumps with a reliable power source will keep heat flowing in the system, to assist laboring thermosiphon systems or to increase the flow whenever collector temperatures soar above predetermined values. Check valves will prevent reverse flow, and freeze alarms can dump primary collector water. Automate as many controls as possible; manual ones should be equipped with audio or visual devices (alarms or gauges) to signal the operator when abnormal conditions exist. Again, keep it simple—there's less to go wrong.

15. Cost. Right at the end of the list, where it should be. When tallied, it may seem high, but you won't think so later on after years of trouble-free service and sky-high fuel costs. It's an investment and, in some states, a partial tax credit—up to 55% in California and other states.

3. Wind Energy

At least once in our lives, we become aware of the wind's power. This may come from studying, reading, or watching a machine convert the wind's energy into pumping water or generating electricity. Or it may come from being buffeted about by a particularly gutsy (or gusty) wind. Whether by typhoon, hurricane, or tornado, many people have seen the sheer power of moving air. The earliest recollection that I have of a demonstration of the wind's capacity was a picture I saw as a child. During a tornado, a straw was driven through a tree trunk 3 inches in diameter —undeniable Krietaslam.

Successful attempts to harness the wind's energy may be seen throughout history; its present use is limited only because of our thirst for energy in more readily-usable forms. Since sources of such energy are no longer so plentiful, the serenity of wind machines and of other alternate energy devices will return. Learning the basics now will prevent a rejection later caused by ignorance. Wind energy utilization is practical, but most people either vastly overestimate or underestimate its complexity. I like to think that it's "involved" rather than "complex," since complexity suggests high technology or hard study before usage is possible. Indeed, water-pumping windmachines are so simple and rugged that no design change has occurred in them for over fifty years.

While wind can be put to work doing many things, this chapter will deal primarily with water pumping and electricity generating. Other tasks, such as the production of mechanical energy, compressed air, or heat, and water dissociation (into hydrogen and oxygen), are viable uses which should be explored further if you find that wind is one of the best alternatives for you.

First, let's get our terms straight. "Windmachine" isn't that bad a word to use to describe the devices that extract energy from the wind, but the term can be confused with machines which "make" wind, such as those which keep frost from settling in the orange groves. A better word might be "aeroturbine." "Windmill" is frequently used to describe anything that sits up there and is turned by the wind, but it correctly refers only to the Dutch units—wind-powered grain-grinding mills. We might consider simply "water pumper" for a water-pumping windmachine. A windmachine that produces electricity could be referred to as a "wind-electric machine," a "windplant," or an "aeroelectric machine," or an "aeroelectric turbine." But let's keep it simple and useful; sometimes it's important to have someone else understand you as well as yourself!

A few novel ideas have been introduced for extracting energy from the wind without using mechanical energy as a step, but for the time being that's the state of the art and that's where we'll begin.

WIND CHARACTERISTICS

I'm tempted to give you the formula for finding the power in the wind, cold turkey, but you should first understand a little about the characteristics of wind.

The wind is, simply enough, a moving mass of air; what it lacks in density, it more than makes up for in speed. Put a windmachine in its way, and the wind will spin it. This means that the windmachine is "gathering" some of the wind's energy—but at a price, of course: the wind is slowed down. Observations and calculations predict that only 60% of the wind's energy can be extracted without adversely affecting performance. Extract it all, and the wind will completely stop behind the windmachine, creating a dead air space; the wind in front of the windmachine will "sense" this pile-up, and go wide around it. This will stop the windmachine, spoiling our afternoon. So, enough energy must be left in the wind to allow it to move on.

Wind rarely blows steadily over a period of time, except in some areas with large bodies of water

Fig. 3-1. A modest wind-electric generator (200 watt Winco) charges batteries for lighting in this New Mexico pit greenhouse.

nearby; offshore breezes and trade winds are known for being more constant and reliable. To the casual observer, there may seem to be little pattern to the wind, and this may make wind energy utilization appear to be a risky undertaking. In years of data measurement and recording, however, distinct patterns have emerged in both wind directions and velocities; annual, monthly, and even weekly patterns occur in some areas. One of the most interesting patterns shows that in many areas, the windiest months are in the midst of winter, and the calmest in summer, directly opposing minimum and maximum solar influx; wind and solar energy complement each other nicely!

Another pattern that emerges indicates that there are two distinct types of wind. The first type is called "prevalent winds," since they blow most of the time and "prevail" over the second type, referred to as "energy winds." Energy winds are commonly known as "gusts" which piggyback the prevalent winds, and deviate from the prevalent wind's direction by as much as 15° to 60°. (A wind vane, then, will "hunt" the wind's direction and, in energy winds, seem unable to make up its mind.) In an average week, we will get five days of prevalent winds—which rarely exceed 15 mph—and two days of energy winds, which range between 15 and 30 mph. The energy winds, though they may blow only 30% of the time, contain 70% of the energy that can be extracted in a week's time! This may seem a contradiction, but it's true, and you'll soon see why.

HOW MUCH ENERGY?

An aeroturbine can extract energy only from whatever wind comes into contact with it. In the same wind, a small windmachine will gather less energy and produce less power than a larger one. It should make sense that in a higher windspeed, both machines will get more energy and produce more power. By further reasoning, it should not seem strange that a small aeroturbine in high winds can produce as much power as a large aeroturbine in low winds. You'll have to take my word for it for now, but some windmachines are more effective than others in converting the wind's energy into mechanical energy (which would best describe the energy of a rotating aeroturbine); we call this "efficiency."

How much energy we get from the wind is related to the efficiency of the machine, its size, the windspeed, and the air's density. These being the only factors, we can now establish a formula that will, once we've plugged in the variables, give us the amount of power we'll get.

(1) $P = \frac{1}{2}\varrho A V^3 E$

Where: P = power
   ϱ (pronounced rho) = density of air
   A = area of the aeroturbine, in silhouette or πr^2
   V = velocity, or windspeed
   E = efficiency

A very intimidating formula, no? Written out, it says the wind's power is equal to one-half the air density times the area of the aeroturbine times the wind's velocity cubed times the efficiency of the aeroturbine. Since air-density variations have negligible effect on power output, we'll give it a unity value (1) and pretend we're at sea level. (If this fudging bugs you, understand two things. One, we have to put air density into the formula to get the units to cancel correctly. Second, the air-density variation between sea level and 10,000 feet has less effect on the formula than a little moisture in the air after a rainstorm.)

Now we can trim the formula down a bit.

(2) $P = .0030685 \, A V^3 (E)$

with: Power in watts
   Area in square feet
   Velocity of the wind in mph
   Efficiency in percentage of 100.

Two things will further reduce this formula to a working size. The first deals with efficiency. I've provided a table (see Section C of Data Cubbyhole) to help you change the formula's number (.0030685) if you'd like to combine the aeroturbine's efficiency with it. At this stage, I've kept it separate for calculation's sake. Notice that the table does not go above 60% efficiency; this recognizes the upper limit of energy that can be extracted from the wind. Rarely does any aeroturbine achieve, much less exceed, this value. I will assign approximate values of efficiency for the different aeroturbines in the section in which they're introduced and in Data Cubbyhole, Section D.

Aeroturbine "area" should be further explained. Irrespective of the "static" silhouette of any type of aeroturbine, its "area" is defined here as its dynamic "sweep." Therefore, you do *not* measure the area of each blade in a propeller-type aeroturbine for use in the formula; rather you calculate its area as you would that of a circle. The length of the blade, tip to

Fig. 3–2. Silhouetted against the evening sky, the author makes a few adjustments.

tip, is the diameter; using the formula for a circle's area (πr^2), you take one-half the blade length (making it a radius), square that value, and multiply it times 3.14 (π) to get the answer in square feet. A 12-foot propeller, for example, is expressed as $(6)^2 \times 3.14$, or 113 ft² of swept area.

If you've ever seen the skinny propellers on a wind-electric machine, you may wonder about those gaping spaces between them and all the wind that must escape. But when the blade is operating at its design rpm, there *aren't* any holes; the blades are sweeping that area so fast that they get almost all of the wind that's trying to go through. If one machine doesn't get as much as another type, that's a matter of "efficiency," and that figure, if you'll look back at the formula, is expressed as E, separately from A, or the area.

Check out the examples in Section E of the Data Cubbyhole to compute the power in wind for varying windspeeds, sizes, and efficiencies, using this formula. After a few tries, either with a calculator or a pencil, you should have no trouble working this out in the future. If you can remember the answer for a particular set of circumstances—size of machine, windspeed, efficiency, etc.—you can easily vary one of these factors by conversion. A 20%

efficiency rating has one-half the power capacity as a 40% efficiency; no need to rework the formula. Be careful, though, when making windspeed adjustments; you're working with a cubed value. While 10 mph and 15 mph can be added nicely to obtain 25 mph, the cube of 10 mph added to the cube of 15 mph does *not* add up to the cube of 25 mph.

THE CUBE LAW

Speaking of cubes, you might as well become aware of why it is so difficult to figure out how much useful energy any system will gather from the wind. The formula tells us to cube the velocity, if we want to find an answer. If we state this another way, we come up with this sobering piece of news: If we double the windspeed, we get *eight times* as much power. You can prove this to yourself by running two examples through the formula, but since that's the only thing that changes, it's simpler just to work with the expression for windspeed. That is, if we cube V, we get V^3. If we cube 2V (or twice the velocity), we get $(2)^3(V)^3$ or $8V^3$. The difference between V^3 and $8V^3$ is the multiplier 8.

Everything else in the formula is *not* a cube value, nor a square value; we've got just plain old multipliers. Believe it or not, we have come to a conclusion. If we want to make leaping increases in power output for small increases in any one factor, let it be windspeed. If we get an eightfold increase in power output by going from, say, 10 mph windspeed to 20 mph, and there's only a difference of 10 mph involved, it shouldn't be too tough to see that if we could just increase the windspeed by 2 or 3 miles per hour, say to 13 mph, we will have *doubled* the power we're getting at 10 mph. We'd have to double the swept area of the aeroturbine or double its efficiency to achieve the same effect as a (calculated) 2.6 mph increase in windspeed!

Since the cube law holds up and down the scale of windspeeds, aeroturbines are designed for higher windspeeds, because that's where all the energy is! No wonder a 100-mph wind is so destructive—it has 1,000 times the power of a 10-mph wind! Most aeroturbines are just *beginning* to deliver some power at 10 mph, and the amount is usually less than 100 watts. At 20 mph, then, this aeroturbine will be generating 800 watts minimum, and at 30 mph more than 2,700 watts is available. The increase in power between 10 and 20 mph (700 watts) is quite small compared with the increase in power between 20 and 30 mph (1,900 watts). The higher winds have it.

Two other effects of the cube law are worth noting. If a weather-data-gathering station is located near you, it has probably computed the average annual windspeed (AAW) for your area; this can vary between 2 and 15 mph. While they only sample the wind velocities at regular intervals, they average these out (including zero wind recordings, too) and you get the AAW. And if you've been paying attention, you're asking if 8 mph is going to be useful to you, correct? Well, rest assured; an 8 mph AAW is good. Note that they average the wind velocity itself; if they averaged the cubes of these windspeeds, you'd get a figure which would give you more real information about the power available in the AAW. Looked at another way, to get an average AAW of 8 mph over a period of one year means that you'd have to have higher (than 8 mph) windspeeds of significant value (or duration) to balance out all those zeros (dead calms).

Still unresolved, however, is the amount of energy we can expect to get from the wind in a year. Certainly, the average annual energy (AAE) is more difficult to compute than the AAW. You cannot calculate the AAE from the AAW. You might get a better answer by computing the AAE from the daily windspeed readings used to find the AAW, but this is still very little information on which to base an expensive system. Installing your own instrumentation, or renting one of the systems available to compute the AAE in kwh per month or year, may resolve the issue. At this point, however, you've become a "digiter," so involved in computations that you've put off doing anything with your hands—like putting up a windmachine and simply *using* it as a way of answering any questions that could possibly need answering. You don't have to buy the biggest and the best at first because, as you'll come to realize, two windmachines make more sense than just one—even if the combined wattages are identical.

ENERGY CONVERSION

There's little that a spinning aeroturbine can do for us as is. But you could connect a shaft to it and, by suitable linkage, have it wash your clothes. Or you could connect a generator to it and produce some electricity. These are two entirely different processes; the first is shifting the energy around, and the second is converting it into another form. It's unfortunately true that with each shift or conversion, we unintentionally lose a little bit of energy to undesirable conversions—heat from friction, noise, vibration, etc. It behooves us, then, to limit the number of shifts or conversions and to minimize, by quality materials or processes, the toll we must pay.

Let's examine the path of energy in a typical wind-electric system (see Fig. 3-3). The aeroturbine (1) converts the wind's kinetic energy into mechanical energy. If used, the gearing system (2) shifts the energy (meanwhile, converting torque to speed) to the generator which (3) converts it into electrical power (horsepower to watts). Electrical wires (4) shift

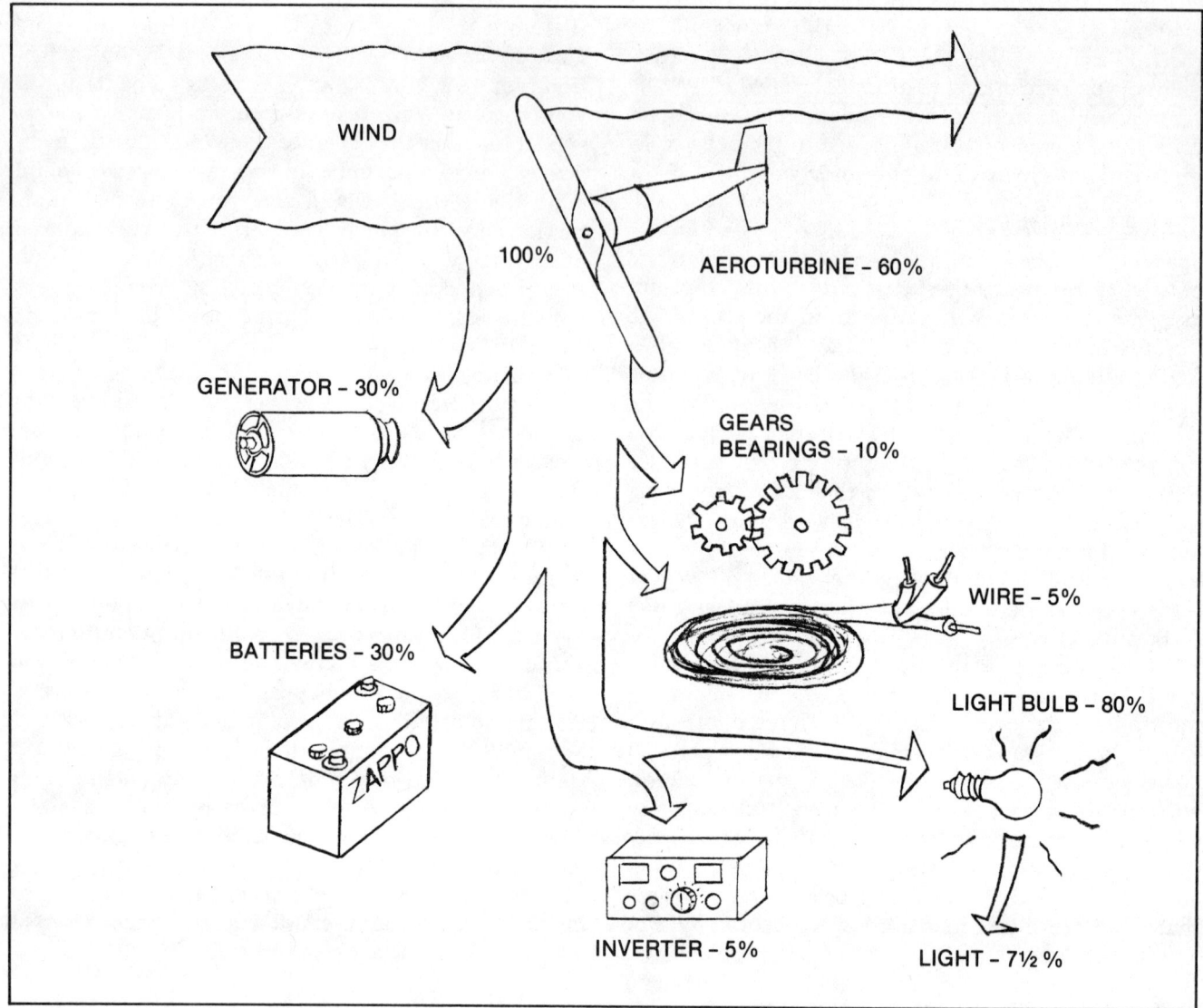

Fig. 3-3. The energy path in a wind-electric system

power from the generator to the batteries, where the electrical energy (5) becomes chemical energy. On demand, the batters (6) convert the chemical energy back into electrical energy which then (7) shifts to the inverter and (8) converts into 110-volts AC, 60 cycles. Finally, it reaches a lightbulb and (9) converts into heat and light.

In these six conversions and three shifts of energy, we have at least nine places where small amounts of energy are lost. Even if we forget that the aeroturbine is not particularly efficient at extracting energy from the wind, the losses add up quickly. The bearing losses at the aeroturbine, gearbox, and generator might exceed 10%. Gearing losses are easily 5%, generator losses (horsepower to watts conversion) would be a conservative 30%, and heating up the wiring soaks up another 5%. Batteries eat 30% of the energy and, with a theft of 5% in the inverter and a whopping loss of 80% in the lightbulb (it's obviously a much better heater than it is a light source), a little bit gets home.

Before you start adding together these percentage losses, stop! You'll only get a horrible mess of an answer that is utterly meaningless. First, convert the *in*efficiencies into efficiencies by subtracting these values from 100%. I get 90% through the bearings, 95% by the gearing, 70% out of the generator, 95% from the wires, 70% passing the batteries, 95% escaping the inverter, and a low 20% that's useful from the lightbulb. Now, don't add these efficiencies. Multiply them! I came out with 7.5% of the original energy remaining or, with an original "gathering" of 1,000 watts of energy from the wind, we're left with a piddling 75 watts. Now that's taxation!

This should help you realize a few things. First, there's no part of the system that can be inefficient without making the whole system inefficient; note that the final "efficiency" of 7.5% is *less* than the least efficient part of the system—the lightbulb—at 20%. Conversely, increasing the efficiency at any part will increase the efficiency of the whole system; this will be more evident if you work with the lightbulb than if you sqeeze more oil into the bearings. If we used fluorescent lights, we'd have an 80% efficient light source, meaning that we'd have 300 watts of light availability. Quite a difference, yes?

We can compensate for the system inefficiencies somewhat by increasing the amount of power available from the aeroturbine. One obvious way is to increase the size of the windmachine so that it can expose itself to a larger section of the wind moving through the area. Or, we can use a type of windmachine that is more efficient at gathering wind than another. But remember when I said that increasing the windspeed even a small portion would double the output? I'll bet you thought I was just speaking figuratively, didn't you? Well, you're wrong! Of course, unless you have a pact with the gods, you can't actually increase the wind's speed. However, you *can* put the aeroturbine where there's more wind. This doesn't mean that you have to move somewhere else; rather, you place the windmachine higher off the ground. You see, wind blows faster higher up in the air than it does close to the ground. While putting your machine on a hill rather than in a valley will help, just putting it higher in any spot will do the same. Trees, hills, and other obstacles (including just plain ol' ground) slow the wind down; putting the aeroturbine up where there are no such obstructions, then, exposes it to wind of a higher velocity.

Before we move on to storing and using the wind's energy, let's finish off the aeroturbine. First a look at different types of aeroturbines, then we'll examine the ratings important to selecting an aeroturbine.

AEROTURBINES

There are two classes of aeroturbine—those with a horizontal axis and those with a vertical axis. The former rotates about a horizontal axis and the latter—yep, you guessed it—rotates about a vertical axis. For many years, the propeller-style horizontal-axis machine (Fig. 3-4) was the only electricity-producing aeroturbine in use. It was able to generate power because it used an airfoil blade. Airfoils are "lift" devices, and they not only travel at multiples of the wind's speed but they extract more useful power. Impulse turbines, like the S-rotor, are essentially "drag" devices, rarely exceeding the wind's speed and producing one-third to one-half the power available from equivalent lift devices. For a long time,

Fig. 3-4. The tail of this 1250-watt Wincharger keeps the single blade faced into the breeze.

none of the vertical-axis machines were lift devices, limiting their value (in the eyes of some people) for producing electricity. This is no longer true for two reasons. The first is that there now exist vertical-axis lift-type aeroturbines. The second is that efficiency is no longer the only criterion for the intrinsic worth of anything. Many other factors are involved in aeroturbine selection for specific applications and they must be given equal, if not more, consideration.

At least two other characteristics distinguish whether an aeroturbine is a vertical axis windplant (VAW) or a horizontal axis windplant (HAW); they are wind orientation and the turntable.

Wind orientation. Windplants in the HAW group must "track" the wind if they are to extract any of its energy. This is accomplished in one of two ways—by affixing a tail (Fig. 3-4) or by running downwind (Fig. 3-5). In the tail version a vane is attached to the windmachine, and the wind pushes this about until the wind pressure is equal on both sides—directly behind the blades. Any wind-direction changes will

66 WIND ENERGY

Fig. 3-5. A windplant designed to run with its blades 'downwind' of the tower

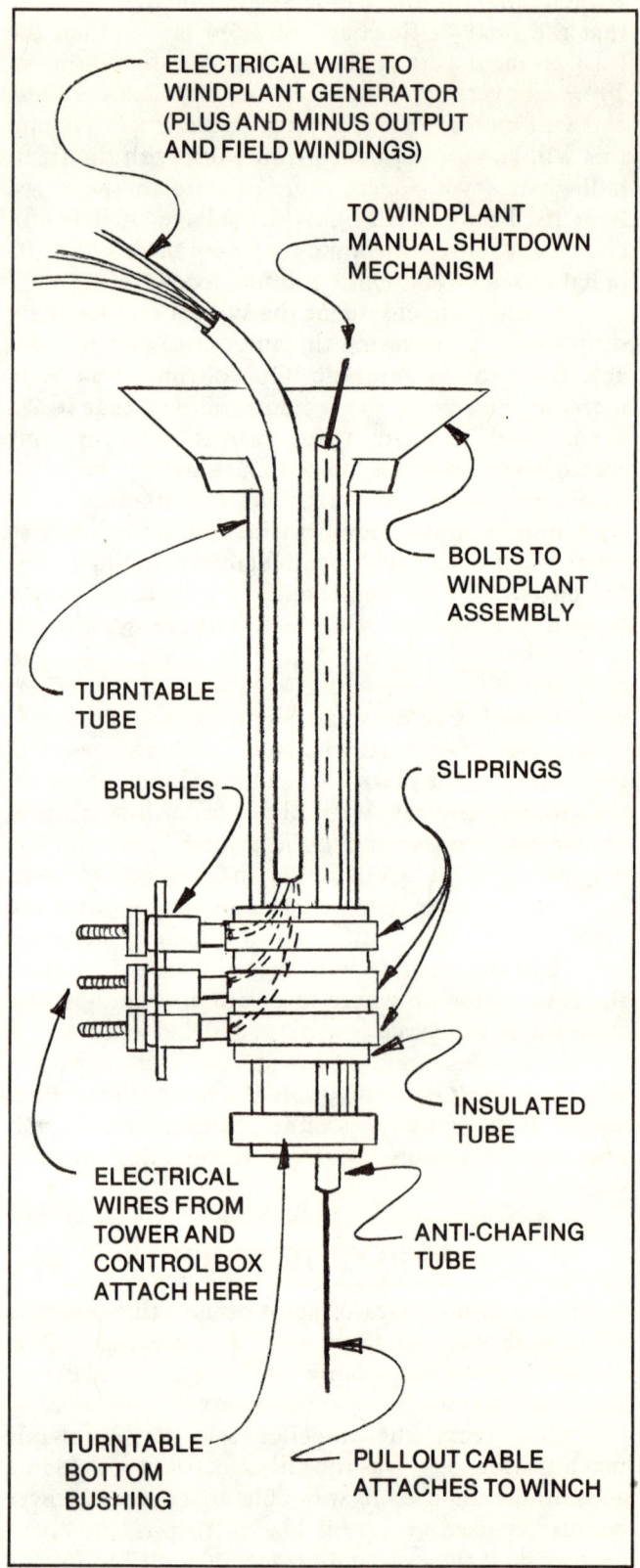

Fig. 3-6. One slipring/turntable assembly design

push the tail the corresponding amount to keep the aeroturbine correctly pointed "into the wind" for normal operation. On the other hand, some windplants are "tailless" or "downwinders." Here, the wind's pressure pushes the machine until the blades are *downwind* of the tower, which is the normal operating position.

The VAWs do not require wind orientation; the foremost advantage of a VAW is that it is always oriented to the wind, irrespective of direction. A better term might be "omnidirectional." The exceptions: some models of S-rotors have "venturi shrouds," and some Darrieus rotors have pitch-changing cams; both require some sort of tail for orientation.

Power will always be lost as a machine hunts the wind, so the VAWs have it over the HAWs there. As well, downwind machines have a slight *dis*advantage over the tailed machines within the HAW group; they're working in the "shadow" of the tower. Unless the tower is streamlined, this means the aeroturbine is constantly exposed to turbulence as the blades sweep from wind into dead air and back into wind; both vibration and structural fatigue of the blades are results.

Turntable. Frequently referred to as the "lolly-

Fig. 3-7. Two- and three-ring slipring assemblies

shaft," the turntable assembly is found only in the HAW; it connects the windplant with the tower's stub. Since either the tailed or the downwinder HAW must orient itself to the wind, the windplant must be free to pivot around the stationary tower; considering that the HAW assembly can weigh as much as 600 pounds (for a unit in the 1,500–2,500-watt range), a hefty bearing assembly is required. That's not too difficult, but two problems arise. How do you get the power from the windplant down the tower? Wires will become impossibly wrapped up and rip apart. And how do you connect the pullout cable for manual shutdown of the windplant? This is the job of the lollyshaft and slipring assembly (Fig. 3-6).

Two methods dominate design in this area; in one the sliprings move about stationary brushes, and in the other the brushes move about stationary rings. In both designs, the pullout cable is routed through the inside of the lollyshaft and attaches to the furling mechanism on the windplant. If the windplant's electrical wiring also uses this tube to connect to the sliprings, another tube is usually inserted in the hollow of the lollyshaft to accommodate the pullout cable and prevent chafing of the electrical wires.

VAW units do not, in normal design, require a lollyshaft or slipring assembly; the generator is bolted directly to the windplant support, and running wires from this point poses no difficulty.

HAW WINDPLANTS

Four types of aeroturbine characterize the HAW group —propeller, windsail, turbine, and dynamo. Nobody else working in the field might recognize this as true, but I'm just using certain words to describe certain windplant designs that are distinct enough from one another to merit examination.

Propeller. The propeller-type HAW reigns supreme for wind-electric generation, boasting some of the highest efficiencies yet achieved from windplants; only recently has it had any real competitors in either the HAW or VAW group. Looking suspiciously like an airplane propeller, this aeroturbine uses two, three, or four airfoil blades to extract energy from the wind. The highest-speed propeller units are the ones with the fewest blades; two blades are faster than three, which are faster than four. Accordingly, a two-blade unit is usually direct-drive, able to achieve the rpm which a standard generator needs to produce power. Four-blade units are usually "ratioed" through gears, chain-and-sprockets, or V-belt and pulleys to the generator; in this manner, the generator is able to turn at several multiples of the aeroturbine rpm. Three-blade aeroturbines are usually ratioed too, but may be direct-drive with a special low-speed generator.

Figuring which to use or build is a windfreak's nightmare. Two-blade units are okay for smaller machines, but shy away from them for more substantial ones; gyroscopic vibration limits their usefulness. This occurs for two reasons. A blade rotating in a vertical plane must fight gravity on the upward stroke and, when going downward, is aided by gravity; it is estimated that 75% of the torque developed in the blade occurs with the gravity-assisted, downward motion. The result is a dynamic imbalance that cannot be adjusted with weights. Another characteristic of a two-blade propeller unit reveals itself when the machine turns to track the wind. The best way to understand this is to suddenly "freeze" the motion with the blades in a perfectly vertical position; if we were to turn the windplant to either side, both blades are equally exposed to the same wind pressure, cutting the same arc at the same time. Now, we unfreeze the motion and then again freeze it, this time with the blades in a horizontal position. The blade closest to the direction of the turn is moving downwind, while the other blade must travel upwind (Fig. 3-8). This can't help but cause some blade-bending, particularly following the "null" experienced with the blades in the vertical position.

Consider what it must be like for one of these blades with the propeller working at 500 rpm (clockwise rotation) and the windplant turning to the right. It's in a null when straight up, downwind when to the right, in a null again when straight down, upwind when to the right, in a null again when straight down, upwind when to the left. That's a 500-cycle vibration, 1,000 changes of direction per minute, and over 16 changes of direction per second. No matter how you look at it, coupled with the torque imbalance previously mentioned, that blade has to take a lot of bump and grind!

Any propeller unit using airfoils is a "lift" device, able to achieve higher speeds than the original windspeed. While it's theoretically possible to achieve

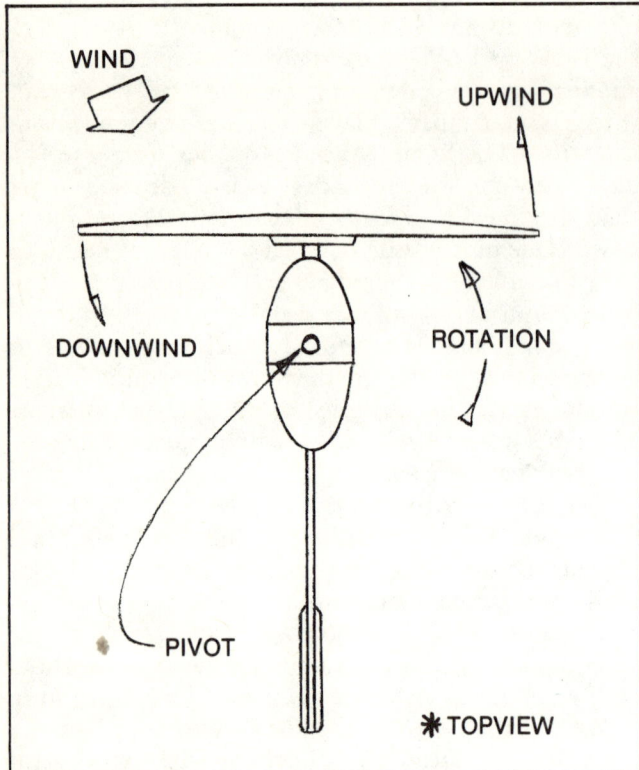

Fig. 3-8. One cause of gyroscopic vibration in a two-blade windmachine

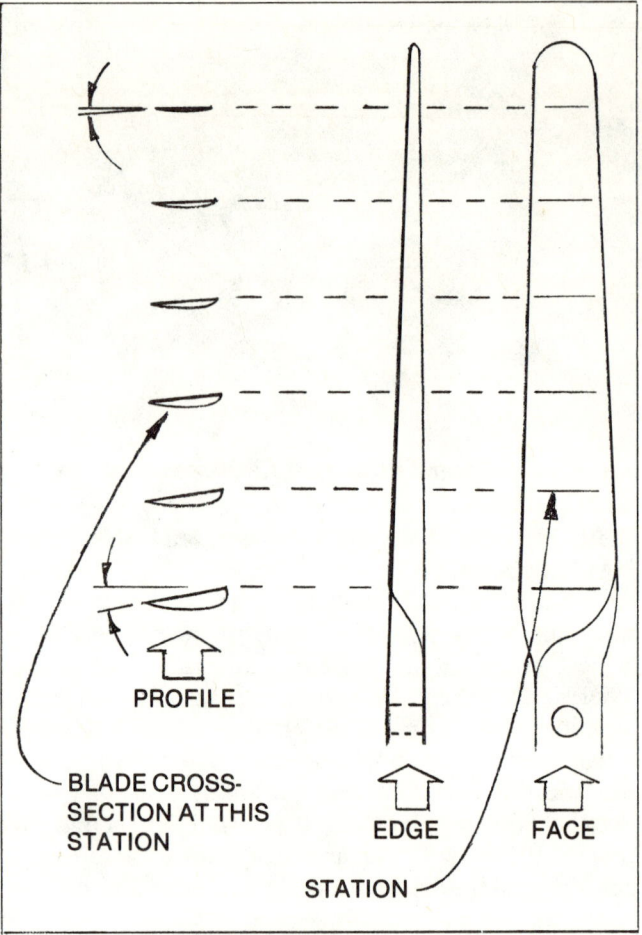

Fig. 3-9. A "taper-twisted" blade

seven to ten times the windspeed, the most practical range seems to be three to five times. This is known as the u/V ratio and is an expression of the propeller's tip speed to the wind's velocity; you use a formula to figure what the rpm would be for a given u/V ratio. Don't lose any sleep over it, though; it's stuff only the real buffs need to know if they're designing a blade. The rest of us knaves just purchase blades.

Three blades, incidentally, eliminate the gyroscopic vibration problem associated with two-blade units. Getting three blades accurately equidistant (120° apart) is not easy, though, and it's one of the chief reasons why four-blade units exist. Proponents for three-bladers don't get along very well with proponents of four-blade units (like the contest between CB'ers and amateur radio folks); the former think four blades is too many. But the truth must be told. Four-blade units *are* easier to build, and in areas of low average annual windspeed, the four-blade units outperform the best of the three-blade units. (Can you tell which group I belong to?)

Another raging controversy in the propeller crowd is that regarding the taper-twist blade. Theoretically (and there can be no denying the math), a windplant will perform better if all portions of the blade become efficient at the design rpm or rated windspeed. If you've ever played in a human whip on ice, there's a certain thrill to being out there at the end, but it is awfully fast. Proportionally, as you move away from a propeller hub, the blade sections are moving faster for any given rpm; the hub may be doing lazy circles when the tip is on its way to breaking the sound barrier. So it should make sense that the airfoil near the tip should be different from the one near the hub if we want both parts to be efficient at the same time (rpm). Consequently, the tip airfoil is thin and narrow and has little angle of attack while near the hub, the blade is thick and wide and has a large angle of attack (Fig. 3-9).

Okay, so what's the big fuss about? Well, some windplant manufacturers a long time ago faced a dilemma. If you have to taper and twist the blade to get this beautiful efficiency, it's going to be expensive. However if you pick one airfoil shape (thickness, width, and angle of attack) and run that the length of the blade, you have a simple and cheap way to make blades (Fig. 3-10). The taper-twist gang adds, "which is inefficient as hell." And so the math would say, too. But some of those who've bought the "improved" taper-twist blades for machines that used to use nontaper, nontwist blades are screaming about poor performance, and they eventually go back to the original blade. This may be exaggerated or it may be the truth; no one personally knows (although many from each side would not admit to that). The poor performance may be due to mismatch; blades which are designed for a rated rpm or generator-loading curve cannot be expected to work well on a windplant which has different ratings than these. A factor left out of many blade-design calculations is the nature of wind itself; test results from wind-tunnel tests should not be generalized to the dynamic performance experienced in real wind.

My own experience (with the usual reservations that this is a feeling rather than documented fact) is that nontaper, nontwist blades of the E-curve type (the standard) take sidewinder gusts better than the taper-twist blades. The E-curve blades seem able to extract energy from these gusts, while the others actually slow down. It should be remembered, as well, that a blade which is designed to become efficient along its length at a given rpm will be the least efficient when at rest. I know—there's no power available in zero wind, right? But to get from zero rpm to the rated rpm means that the blade must be able to respond to wind of much lower velocities, and there can be no argument that a blade that's nontwisted and nontapered fills that bill. Only side-by-side wind plants—one with taper-twist blades, one with nontaper, nontwist blades—of identical size, the same number of blades, at the same height, and exposed to the same wind will verify or refute any claim made for either blade design. Sounds like a great grant proposal, if only money were given to really worthwhile projects!

Since propeller-type aeroturbines reach the highest speeds (rpm *and* blade tip velocities), they automatically get the vote for the most dangerous. These blades must be strong yet light—the best combination for dealing with centrifugal forces.

Windsail. A second type of aeroturbine of the HAW group is the windsail; this is the gentle version of the propeller units. The windsail uses between four and eight sails rigged from booms, similar to the old Dutch windmills (Fig. 3-11). Its sails are typically made of cloth (nylon or canvas). By ingenious rigging, the sails can simulate an airfoil shape and, consequently, achieve some of its advantages, but for the most part they're "drag" devices, achieving rotational speeds equal to the wind's speed. The sails, when spread to the maximum, present a large surface area to the wind; as a result, they're very happy developing high torque from low windspeeds. This makes them ideal for use in water-pumping applications, but creates problems in generating electricity. For overspeed control or adjustment for best performance in the winds characteristic of any given area, the sails may be furled (rolled onto the booms), presenting

Fig. 3-10. This 1500-watt restored Wincharger sports four aluminum non-tapered, non-twisted blades.

Fig. 3-11. Unfurling the sails on a Windsail aeroturbine *(Courtesy James L. Ruhle & Assoc.)*

Fig. 3-12. The American Wind Turbine uses a rim drive arrangement. *(Courtesy James L. Ruhle & Assoc.)*

less surface to the wind. Or how about an automatic furling device? No better system protects the aeroturbine in high winds than one in which the aeroturbine virtually disappears.

Turbine. The turbine is the type of HAW which has been used for many years on farms throughout the United States for pumping water; if you're scratching your head, it's what everybody calls a windmill when it's pointed out. Having more vanes than the windsail has sails, this aeroturbine does all its work in windspeeds between 5 and 15 mph. The first practical application of the turbine to the generation of electricity occurred in the early 1970s when airfoils were attached to the spokes of a wire wheel 15 feet in diameter (fig. 3-12). In tests, this 50-blade aeroturbine dispelled once and for all the theory that too many blades got in each other's way and that a machine using that many could not produce electricity. By attaching a rim at the blade tips, a second problem associated with low rpm aeroturbines and high-speed generator combinations was solved: A ratio drive was not necessary when the generator

could "ride" on the rim of the aeroturbine, where surface speeds were high.

Dynamo. Not too many sections back, I was discussing a number of ways to get more power from a certain size aeroturbine. Two solutions were presented in recognition of the cube law of wind (the power in the wind goes up with the cube of the velocity, remember?). If you can't increase the wind's velocity (that's playing God), you can put the windmachine where there is more wind (higher in the air), or use a more efficient aeroturbine. Well, I lied. You can, within limits of course, play God when it comes to wind energy, for there is a way to increase the wind's velocity: use a venturi. Theoretically, this shroud doubles the air's velocity and halves the density; if we plug that into the wind formula ($P = 1/2 \varrho A V^3 E$), we'd see four times the power availability from the same wind. In practice, there are some serious difficulties with this, particularly with the formation of a lot of dead air behind the blades; whether this design will ever be practical or not is controversial. However, again in the early 1970s, an energizing young man put together a dynamo: an airfoil-shaped, venturi-shrouded aeroturbine (Fig. 3-13). Besides using the theoretical advantage of the venturi, this man put magnets in the blade tips and the stator coils in the venturi shroud. To the layperson, that means he built a windplant that didn't *turn* a generator, but *was* a generator. For reasons unknown—lack of funding, other technical problems, etc.?—little else has been heard of the dynamo in this decade. For the experimenter in us, this is an intriguing concept, but perhaps one that is out of reach.

VAW WINDPLANTS

For many years, there was only one aeroturbine that fit into the VAW group—the Savonius rotor. But there are two more recent additions, the Darrieus rotor and the hybrid. Their ability to face 360° of wind is their distinct advantage over the HAW group; while the others hunt the energy-bearing wind or split the difference between winds from two directions, the VAW group is merrily taking advantage of anything that comes in, from any direction. VAWs, as well, normally operate at lower rpm, lessening the need for superstrong materials or finely balanced aeroturbines which will survive exposure to high rotational speeds.

S-rotor. Also known as the S-rotor, the Savonius rotor (named after the inventor in the early 1900s) is a VAW "drag" device; its rotational speed will usually not exceed the wind's velocity. While extensive research followed its discovery, it was used just for some water-pumping applications; the author is credited with the first practical application of the S-rotor to generating electricity. In simplest form, the

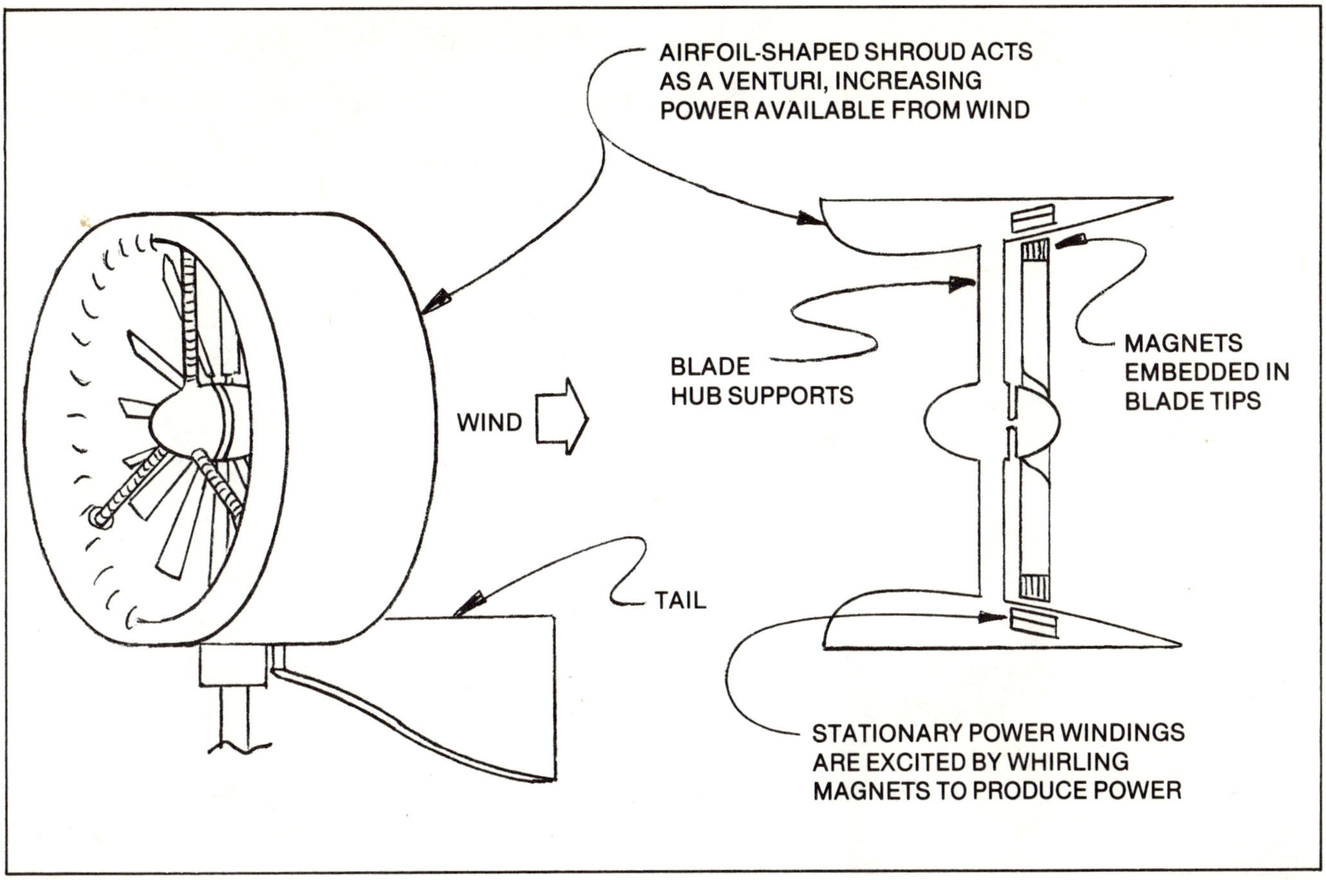

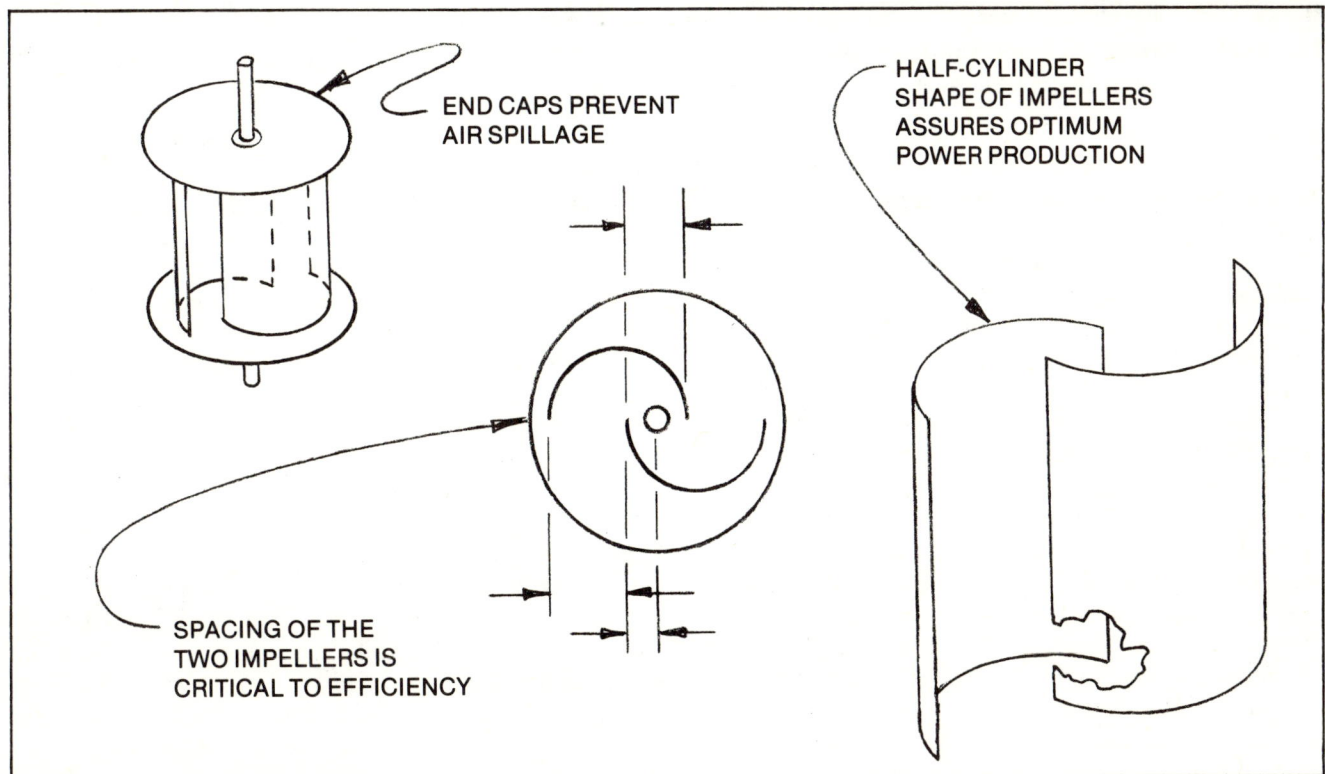

Fig. 3-13. The dynamo: A venturi-shrouded propeller windplant

Fig. 3-14. An S-rotor

Fig. 3-15. Earthmind's homebuilt three-tier aluminum S-rotor and support framework

Fig. 3-16. An experimental Darrieus rotor. *(Courtesy James L. Ruhle & Assoc.)*

S-rotor is a cylinder which has been split through its length into equal halves; these halves are then offset by a distance equal to approximately one-half the original radius, and secured by endplates which are larger than the new diameter (Fig. 3-14). Since a starting problem exists if the S-rotor comes to rest in a particular position, single S-rotors are usually stacked in tiers out of phase with one another (Fig. 3-15).

The use of simple tools, readily available materials, and minimal construction skills all constitute advantages favoring the S-rotor, particularly for the owner-builder. Methods now exist for mounting the S-rotor on top of, and inside, towers (this was a serious problem only ten years ago). Because it exposes a large frontal area to the wind, the S-rotor develops torque rapidly as the wind speed increases. However, since a portion of the aeroturbine is always moving upwind (for no gain in power), its physical size must be quite large to match the power from a propeller unit. For this reason, and other problems experienced in scaling up any windmachine, users often build and mount several S-rotors of smaller size to equal the output of a single large unit.

D-rotor. To refute the alleged fact that VAWs could not compete in efficiency with the HAWs, the Darrieus rotor was conceived. Affectionately named the "eggbeater" for its unorthodox appearance, this device mated an airfoil blade with the VAW to produce a formidable two- or three-blade windmachine (see Fig. 3-16). Empirical observations quickly revealed two speeds at which the D-rotor would operate. At the higher speed, the blades actually sweep through the same wind twice—once as the air initially hits the machine, and a second time as it exits the aeroturbine downwind.

Several innovations with the D-rotor have kept it alive in the wind-energy field. One uses a high-aspect tail to hunt and find the predominant wind direction. This is tied to a "follower" cam which changes the pitch of the blades as they move in rotation, biasing the blades to take best advantage of the relative wind irrespective of their position at any given moment (Fig. 3-17). In this and other versions

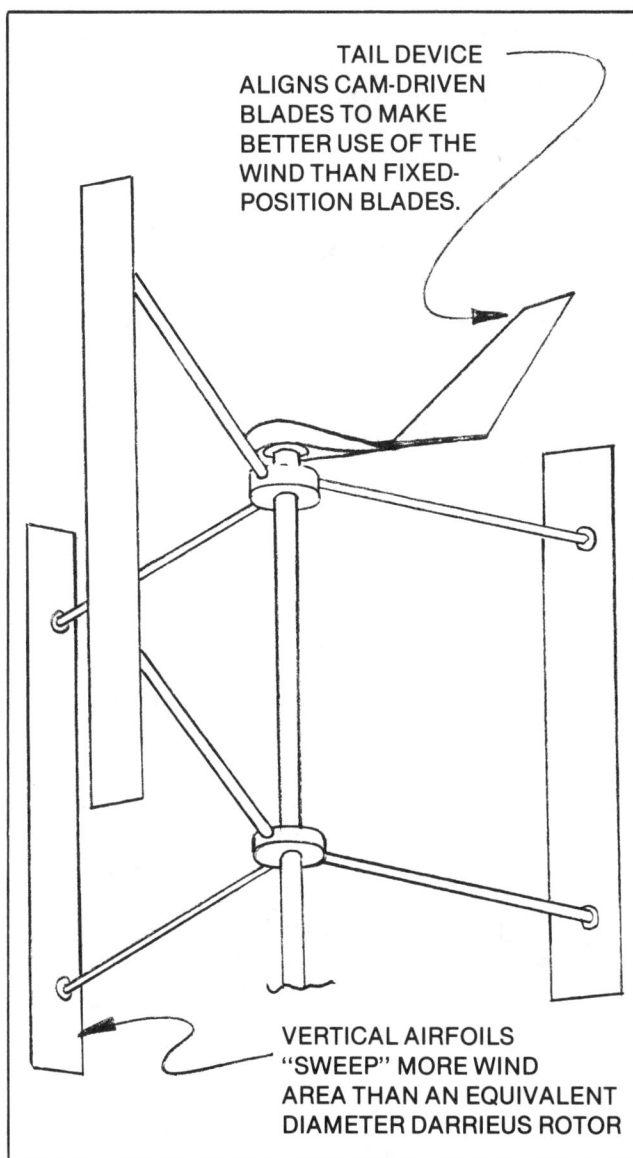

Fig. 3-17. The Pinson Cycloturbine: A uni-directional, 3-bladed hybrid

Fig. 3-18. A hybrid windmachine (S-rotors in center of modified Darrieus rotor)

of the D-rotor, the blades are vertical, so more area is swept (and more wind gathered) than the "egg-beater" style.

Hybrid. The hybrid VAW is a combination Savonius rotor and Darrieus rotor, initially joined to combine the best features of each. Pure D-rotors have serious starting problems, needing windspeeds higher than their "cut-in" windspeed to get moving from rest. Pure S-rotors are limited, as aforementioned, in their power-producing capability. Mix a high-starting-torque, drag-type S-rotor with a high-efficiency, lift-type D-rotor and you have a hybrid (Fig. 3-18). The final scores aren't in yet, but it doesn't appear to have worked; the hybrid starts, but doesn't perform well. At present, it's back to the pure D-rotor which gets an initial kick from its own generator being "motored" when the wind reaches power-producing speeds. And a few people (including this author) are conducting tests by modifying the S-rotor for better performance.

AEROTURBINE CLAIMS

There is no reason for anyone to get burned buying a windplant. With the power formula for wind in your head, your calculator in one hand, and a notepad and pencil in the other, you're ready to go look at windmachines. When someone tells you that his or her machine delivers so much wattage at such and such a windspeed, ask that person to wait just a minute. Write out the formula on the pad. (If you've done this beforehand, it doesn't have quite the same effect on the salesperson.) Figure the swept area of the machine you're checking out and enter this in the formula. Take the power figure the seller gives you and double it; since you're losing about 50% of the power getting it through the generator, the aeroturbine should produce at least twice the true wattage the windplant produces. Enter this figure in the formula. Take the rated windspeed figure, and cube it. If it's 25 mph, for example, that's 25 × 25 × 25, or 15,265. Now, solve for E, or efficiency. If it comes to over 60%, the salesperson should get the Nobel Prize for doing the impossible. Stroke a finger pointed at them with a finger from the other hand, and say, "Shame, shame."

If the efficiency figure you get is something between 30% and 55% and you're looking at a propeller or turbine unit, it's probably okay. Still, be skeptical of figures toward the higher end. A Darrieus or Savonius rotor won't be found in this efficiency range; they're more like 10% to 30%, if that high.

You'd be surprised (or maybe you wouldn't) at the number of reputable-looking firms that have beautiful windmachines to sell which can't or won't live up to the claims made for them. If some manufacturers used the windpower formula, they'd realize that their machines cannot do what they say. Okay, maybe they tried and dropped a decimal point; either way, it's apparent that the machines have never been exposed to real wind and delivered a certain amount of power in the desired windspeed. Which means that windmachines are being rated, in general, just as cars are rated nowadays for gas mileage—by mixing engine and power-train efficiency curves, on the best road conditions, for the best driver, driving his best—giving us the most meaningless figure we'd ever want. Use the formula they've abused and get an answer that meets with the facts, or do business with someone else. It's hard enough getting the wind to fill your sails without having holes in them too.

Don't forget the earlier comment in favor of several windmachines instead of just one. If they're identical, you can always cannibalize parts to keep at least one of them going, and, for buying spare parts, you might get a price break buying more than one. Two windmachines doesn't mean double of everything else, either; towers designed to carry two windmachines aren't all that difficult or rare. Each will need its own control box, but ammeters and voltmeters and other hardware can be shared with a simple switch reading one or the other. Both can charge the same battery bank provided they're identical in the rated voltage. I realize that contemplating two windplants is difficult when you haven't even decided on one yet, but it's a good thing. Ever hear of a music buff buying only one speaker? Yep, stereo windplants!

WINDPLANT RATINGS

All aeroturbines, irrespective of size or type, have lower and upper limits (usually expressed as particular windspeeds) below which they will not produce any power and above which no increase in power occurs. Below the lower limit, called cut-in, the aeroturbine is stationary or moving too slow to be effective; power is first developed around a windspeed of 7–10 mph. Above the upper limit, usually referred to as the "rated" or "maximum" windspeed, the machine is developing its designed power level. Although it would be lovely, indeed, to gather the energy that is so abundant in these higher windspeeds, we should not get greedy. It is now time to concern ourselves with the aeroturbine's safety, since it is possible for it to go too fast and destroy itself. Whatever automatic governor is designed into this system will not come into action, limiting the aeroturbine's rpm and, consequently, its power output. The windspeed may increase and the aeroturbine still deliver power, but it will (or should) not exceed the amount developed at its "rated windspeed." So, if it normally produces 2,000 watts at 25 mph, it will still produce 2,000 watts at 40 mph. If you're expecting a hurricane, tornado, or typhoon, however, it's best to shut down the windmachine (using one of many techniques); there is a limit to the protection afforded by the automatic governor!

I've taken the liberty of using some words—power rating, windspeed rating, etc.—which I have not fully defined, so let's examine them now. If you're going to buy a windmachine rather than "homebrew" one, these next few sections are meant for you. For the do-it-yourselfers (hereafter referred to as DIYers), read and heed; these are necessary design parameters.

POWER RATING

This is a much-discussed yet little-understood characteristic of windplants. How much energy you need is *not* expressed by the power rating of the windplant. If you want to know if a particular power rating of windplant is going to work for your situation, you must involve yourself in a bit of calculation and some basic soul-searching regarding your own use and waste of energy. There's no rule of thumb here; the size of windmachine for your situation involves several dozen parameters.

If you're buying a wind-electric system, beware. If a salesperson had to write down and sign everything he or she was willing to tell you, there'd be a lot fewer claims made for a particular system. On the other hand, if you go in there with little or no knowledge of the subject, you're not going to understand a manufacturer's claims, even if they are true. You can't blame someone for selling something to a customer who's too ignorant to know the product won't work under the circumstances.

One area of particular concern is power rating. Since there is no real industry for windpower (at the time of writing this book), there are, understandably, no standards. So you ask the question almost everyone else asks first, and that is: how much power does it produce? And you get an answer: 2,000 watts. Now, what does this mean to you? Actually, it shouldn't mean anything. You have another question you should immediately ask: from where? You see, some folks rate their machines by what the aeroturbine produces. If you recall our stroll down the energy path for a wind-electric system, this can be twice the amount actually delivered from the generator. Maybe someone is trying to rip you off and maybe not—but if you don't ask, you'll never know, will you?

WINDSPEED

The next question you should ask after being informed that such-and-such a windmachine will deliver 2,000 watts is: at what windspeed? If you don't, you should have your hand slapped and be sent home with a note. The power rating of a windplant is just so much nonsense unless you know this as well. Many windmachines currently being manufactured are rated to deliver full power at 25 mph windspeed; this rating is useful only for those areas of the world experiencing high AAWs (average annual windspeeds), usually above 12–14 mph. I've noticed that many people in the U.S., for example, purchase these machines from home or abroad when, in fact, they will only rarely develop full power. A windplant that develops its full (rated) power at 18 mph would be a much more suitable machine for 90% of the U.S. Unfortunately, since most machines won't do this, most manufacturers fail to mention the "rated" windspeed at all in their brochures, an action I consider criminal. Many will specify at what windspeed the windplants will begin charging, but this is almost useless information, since there is so little power in low windspeeds anyway. Don't be afraid to ask and, for any answer, *get it in writing!!*

I've mentioned the more desirable windspeed rating to some folks and had them shrug at the 7-mph difference between 18 and 25 mph as though it didn't matter. Suckered 'em right in, I did. Then I asked them to guess how much power, say, a 2,000 watt windplant, rated at 25 mph, would deliver at 18 mph. I got all sorts of guesses, but none of them low enough. Do you know? Well, do the computation with our wind formula or take my word for it; it comes to 746 watts on the nose! So if someone was selling a windplant that delivered 750 watts at 18 mph, that betters the performance of the 2,000 watt windplant at the same windspeed. For the difference in prices asked for each, you'd be better off buying two of the 750-watt models than one at 2,000 watts which only occasionally experiences wind in excess of 18 mph. Another good reason to own several windmachines instead of just one!

Specification sheets for the manufactured windmachines sometimes show power-curve graphs, but many don't; you'd better bring along a calculator to adjust whatever values they do give for other machines, as a means of comparing one with another. Sorry, it's the only way to get the most windmachine for your money. Don't expect the seller to tell you the *dis*advantages of his product; you have to surmise them from what he doesn't say!

STORAGE SYSTEMS

To effectively use wind energy requires some means of storing it for use in nonwindy periods. This means changing the kinetic energy of electricity into a potential energy which can be efficiently reclaimed (as electricity). At this time, the least expensive and most reliable means of doing this is chemically—in a battery. Some people would disagree loudly at this statement, but let's check it out.

AC vs DC

Most windmachines deliver DC (direct current) rather than the common household AC (alternating current). There are a couple of important reasons why. First, you can't store AC in a battery. Second, it's wasteful, expensive, and difficult to try providing the very specialized AC used in the household—60-cycle AC. That doesn't mean, however, that it's not being done; you *can* buy 60-cycle inverter systems. These use the utility line as the storage bank; if you've heard about a way to run your meter backward, the cutie that does it is the synchronous inverter.

Fig. 3-19. In some areas, the utility company requires two meters for use with a synchronous inverter system —one for power you consume, one for power you produce.

While most people will probably want the convenience of this special kind of power—110 volts, 60 cycles, AC—it's full of holes as far as I'm concerned. The synchronous inverter may eliminate a battery bank for storage, but there's a limit to how many people can use the system, and it requires continuous connection to, and dependence upon, the utility system. If there's a blackout, the unit is inoperative even if the wind *is* blowing, which is a very ridiculous situation. Its only advantage lies in the area of commutation—if the area in which you wish to use the energy is far removed from the area in which you must locate the windplant. This is a very complicated and expensive system itself, however.

BATTERY RATINGS

There are four major ratings to consider in batteries for a wind-electric system—capacity, rate of discharge, deep-cycle capability, and system voltage. Plunge ahead.

Capacity. Capacity describes a battery's storage capability. This is *not* rated in kwh (kilowatt-hours). The first thing you should understand is that batteries do not store electricity. Rather, the electricity that goes through a battery during charging causes a chemical reation. Then, when you put a "load" across the battery (turn on a light, for instance), this causes the chemical reaction to reverse, releasing electricity. Temperature, the rate of discharge, and other factors will dictate how much electrical energy you can expect to get out of a battery. So that manufacturers can have some common ground in describing a battery's capacity, they've developed a rating system called the ampere-hour, or amp-hour, or (if you have to write it a dozen times) the ah. The number preceding the ah is the capacity of the battery; 60 ampere-hours or 205 ah are examples.

Rate of discharge. The ah rating is, by itself, useless. A battery which is designed to deliver 3 amperes for twenty hours won't necessarily deliver 20 amperes for three hours, and certainly not 60 amperes for one hour or 120 amps for half an hour. But if you multiply the amperes by the hours for each of these, you'll get 60 ah. So, manufacturers will not only specify the ah rating, but also indicate the "design discharge rate" in hours. Automotive batteries, for instance, are given a twenty-hour rating. To determine the amperage, you need only divide the ah rating by 20, and you'll know the "rate of discharge." The guarantee by the manufacturer, therefore, is that a 60-ah battery will deliver 3 amps for twenty hours. In a car, it's a whole different story; the battery's main purpose is to start the engine, and it certainly *won't* do that at 3 amps—it's more like 200–400 amps. Fortunately, engines start in a few seconds; if it takes longer than a few minutes, we all know

Fig. 3–20. Batteries for a wind-electric system

what happens, don't we?

I've already indicated that a battery won't deliver varied amounts of current at different rates than the specified one, but understand that the general rule is: the higher the discharge rate (relative to the rated discharge), the less ah the battery will deliver. But if the discharge rate is less (or the hours longer) than the specified rating, you can expect to get more ah than the specified amount. This should indicate that an ah rating does not define the true capacity of the battery—just what it will deliver at a particular rate of use.

The battery you want to get for your wind-electric system will have a high rating—probably above 220 ah—but you should get it at a discharge rate of six to eight hours, if possible. Motors and heaters like current, so you have to get a battery that will deliver current without sacrificing some of its capacity in heat losses.

Deep-cycle capability. There is only one type of battery to get for your wind-electric system, and that's the "deep-cycle" variety. The name defines the battery's unique design: an ability to withstand discharge to a very low value, say 15–20% of its capacity, repeatedly, without damage or destruction. Do that to a car battery for a while, and you'll be in the market for a new battery! Deep-cycle batteries, however, will do it as many as 2,000 cycles (each cycle is composed of a charge and discharge) in a wind-electric system. That's about twenty years of good service!

One of the characteristics of the deep-cycle battery is thick plates. Those of an automobile battery are paper-thin, for lots of surface area, giving it surplus amperes when they're needed—to start engines. But a thick-plated battery is not necessarily designed to deep-cycle. Truck batteries are not deep-cycle; neither are most "Die-Hards." Since advertising is always a bit less than accurate, be persistent, and skeptical, and move slowly. The time to replace your

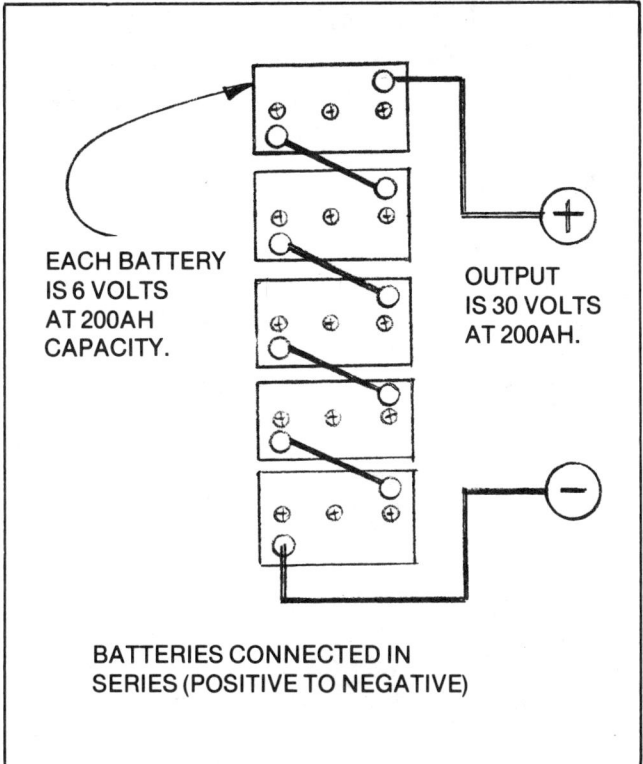

Fig. 3-21. Voltage is additive when batteries are connected in series.

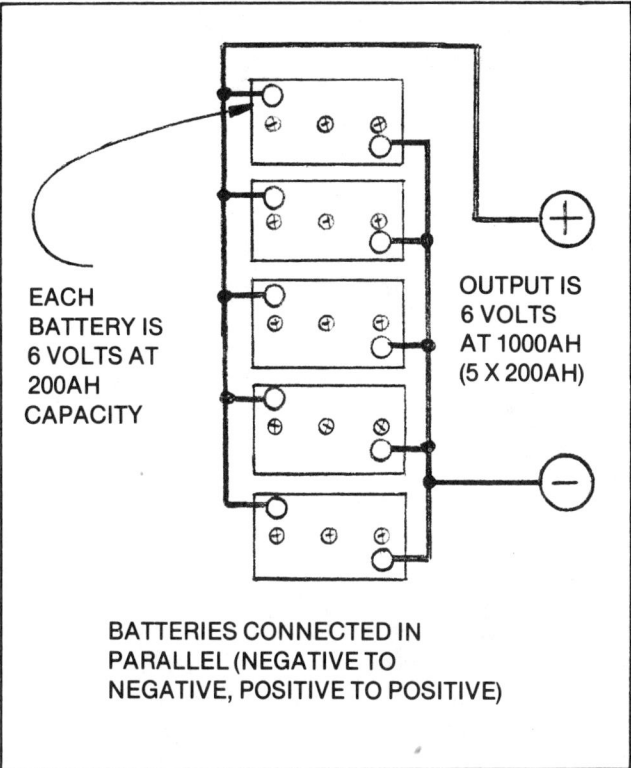

Fig. 3-22. Ampere-hour (capacity) is additive when batteries are paralleled.

batteries will come soon enough (twenty years goes by in a flash); don't rush it by getting anything less than what you need.

Voltage. If you've got a little cabin tucked away somewhere back in the deep woods and you mean to simplify your life, stick up a little 200-watt windmachine for the music, a reading lamp or two, and the rechargeable flashlight. Usually this is a 6- or 12-volt system and, for the small amounts of power produced or consumed, it's most adequate.

For the rest of us, operating motors, tools, appliances, etc., the voltage or the capacity of the system must be increased. So, who's ever heard of a 32-volt battery, huh? There's a tendency to think of a battery as "singular" when, in practice, it is a number of 6- or 12-volt batteries connected together to equal 32 volts (or any other required voltage up to and including 110 volts or 220 volts).

By definition, a battery is two or more cells; a cell is one chamber of lead plates and acid (normally sulfuric acid) which will produce 2.2 volts. A 6-volt battery, then, has three cells and a 12-volt battery has six; since each cell has a combination vent-and-fill cap, three caps means we've got a 6-volt and six caps identifies a 12-volt. You won't usually see batteries with voltages above 12 volts because, if they've sufficient capacity, it takes a forklift to pick 'em up. Typically, a 6-volt 200-ah battery will weigh 60 pounds, as will its equivalent, a 12-volt 100-ah battery. Higher ah ratings are impossibly heavy for a 12-volt or a 6-volt battery; by the time you get to 300 ah or more, you'll be wishing that they made them in 3-volt packages (nope, that's like a $3 bill). They do make them in 2-volt sections, but it would be incorrect to call these batteries; they'd be "cells." But there are alternatives to grouping 2-volt cells, which are much rarer or more expensive.

Since 6-volt and 12-volt batteries do exist and they're composed of nothing more than cells, we may infer that cells and batteries alike may be connected in such a way as to achieve almost any voltage desired (Fig. 3-21). With the plus-to-minus wiring, we will increase the voltage and maintain the ah rating of any single battery; five 6-volt, 200-ah batteries connected "in series" will result in a battery pack of 30 volts at 200 ah. While it is possible to use individual batteries of different voltage to achieve a practical battery pack, each battery should have the same ah rating. Batteries of the same voltage but differing ah ratings should *not* be connected in series; there's no way such a pack can be charged without overcharging or undercharging some of the batteries within the pack.

Another useful arrangement of batteries (Fig. 3-22) is called "parallel" wiring. When two 6-volt, 200-ah batteries are paralleled, the ah ratings become additive and the voltage remains the same; we end up with a 6-volt, 400-ah battery (or battery pack). If we add another in parallel, we get 6 volts at 600 ah of capacity. This is helpful if we don't want to buy 2-volt cells in order to get capacity for the system; paralleling becomes downright necessary if we have a low-voltage windplant and need lots of storage for it. The beauty of parallel packs is that you can add on to your battery storage as use dictates, and the pocketbook permits. And, as long as you keep batteries which are in series of the same ah rating, the parallel packs can be of differing ah capacity; a 30-volt pack of 200 ah paralled with a 30-volt pack of 350 ah capacity is perfectly legitimate.

Through switching or isolation diode configurations, you can charge or discharge battery packs separately or together. This is beneficial where you're not sure how much energy is stored in either pack; use only one pack at a time and the other serves as a backup if you drain the first. This saves on blackouts and brownouts while you're developing a feel for your system. Sure, most wind-electric systems have backup electricity-generating capability (we're coming to that soon) but, by developing a feel, additional storage requirements become evident. Besides, once you've experienced this gentle, quiet, and graceful system, you will detest the intrusion a standby generator makes into your (otherwise) serene life.

The selection of a lower- or higher-voltage system is more circumstance than choice. While it doesn't have to be, the windplant's voltage usually dictates the battery voltages. If you're not given options of voltage for the type of windmachine you wish to purchase or build, the simplest solution is to remain with that voltage. To help you decide whether or not to build or buy any type of system, here are the pros and cons of voltage levels.

High-voltage systems require large quantities of batteries. A 110-volt system needs fifty-six 2-volt cells, nineteen 6-volt batteries, or nine 12-volt batteries. This is roughly four times the minimum number required for a 32-volt system, five times the number required for a 24-volt system, and ten times the number required for a 12-volt system. Only the 12-volt and 110-volt systems, however, have readily available appliances, tools, etc. that operate at these voltages; most households operate 110-volt stuff already and the camper craze has resulted in an abundance of 12-volt gadgets.

High-wattage use of wind-generated electricity calls for higher voltages, or the highest obtainable. Unless you can keep those power-consuming devices close to the battery bank, very large-size wire must be used to carry current from low-voltage systems; this is expensive. Without it, however, too much of the energy gets used up heating the wire. In a new house, at least, cost is the only real factor, and the installation of larger wire is rather easy. In an older house, rewiring provisions can be as expensive as the wire itself; it may be easier to opt for a higher-voltage windplant.

A supreme advantage with lower-voltage systems is a lessened hazard-to-life setup; while lower systems can still kill, it's not nearly as sure a thing as with higher voltages. If you're homebrewing the system or want to become very involved with the maintenance and repair of your manufactured wind-electric system, you'll get shocks no matter how hard you try to avoid them. It's a comfort to know, with lower-voltage systems, that the experience will result in a definite reminder to be careful, rather than your first flying lesson—nonstop—across the room.

Working with your own energy production is a fulfilling pastime. To keep it that way requires a good understanding of battery maintenance. This means adding water to the batteries as required; "gassing" will deplete it over the months. Using a hydrometer (a device for measuring the specific gravity of a battery and a means of determining its state of charge) you can validate the system's correct operation, or pinpoint improper control-box settings. You may have to remove accumulated dust and debris from the batteries, clean and tighten connections, and (if you don't live in a high-wind area) give the batteries a once-a-month equalizing charge, to assure their long life. Whenever working around batteries, you must beware the danger that lurks, for it is as deadly as it is silent. Voltage is one of the greatest dangers, but gas is another. The phenomenon of gassing means that free hydrogen and oxygen are produced during intense charging and discharging of the batteries; it is imperative that adequate ventilation be provided for the battery storage room (or space). Besides the unpleasantness of the "dry throat" you get from the presence of free hydrogen, there's always the excitement of the *Hindenburg* effect; it only takes 4% hydrogen and one match to demonstrate it. If you don't have tight connections at the battery terminals, or you drop a wrench between the posts, you don't even need the match. Even if only one battery (instead of the whole bank) explodes, you've still got a good chance at getting a cardiac arrest from the shotgun noise at close quarters, or some hefty burns from sprayed acid. Carelessness and batteries don't mix—quietly, that is.

AUTOMATIC WINDPLANT CONTROL

Control of a wind-electric system encompasses a wide variety of functions. Were these not automatic, we would spend much of our time monitoring windplant operation. For proper operation and protection of the aeroturbine, generator, and batteries, some form of control must effectively deal with high windspeeds, overcharging, current limiting, etc. This is the job of the control box and the governor. Both function to protect but allow for normal operation, sensing combinations of conditions and knowing what, when, and how to do what should be done.

CONTROL BOX

For all they do, control boxes for wind-electric systems are extremely simple devices; a $10 transistor radio is far more complex. But what does a control box do? Let's list some of the standard tasks it performs. The control box:

1. Allows the generator to begin charging the batteries when it reaches the point (windspeed or rpm) where this is possible; when below this point, it disconnects the generator from the batteries to avoid draining power from the batteries.

2. Protects the windplant generator from putting out too much voltage or current, which can cause arcing or excessive heat in the generator, and possibly fire or severe damage.

3. Senses the condition of the batteries, allowing a high rate of charge when they are near exhaustion (from use), and a limiting action (less current) when the batteries are fully charged and no heavy, power-consuming devices are being operated.

4. By virtue of the action described in #3, protects the batteries from over- and under-charging and, in the instance of the former condition, excessive gassing.

5. Directs the current from the generator to the batteries and the "load" (anything which is consuming energy), or, if the windplant is not operating, from the batteries to the load.

6. Indicates charging current (from the generator) or discharge current (from the batteries) when a load is connected; this can be "read" on one of two meters or, if the meter is plus *and* minus reading, one meter. A voltmeter may be also used for monitoring purposes, reading battery potential or generator voltage.

7. Serves as a junction box for the various components of the wind-electric system; wires from the load outlets, the windplant, the standby generator, and the batteries are all connected at the control box.

8. Houses the meters, relays, diodes, resistors, switches, fuses, and interconnections of these components (Fig. 3-23). Sound almost too simple to say? Not really. The box serves to protect them from dust, moisture, and physical abuse or damage, and to give you only one place to go when there's trouble.

The control box is a brain—switching, routing, and limiting. It houses the controls which can be

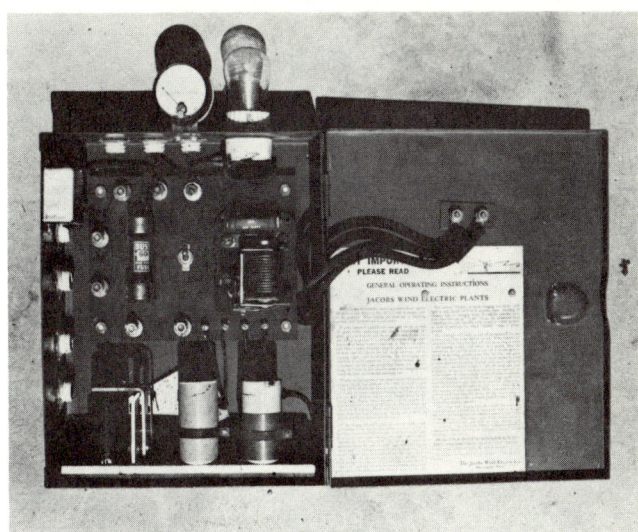

Fig. 3-23. The innards of a Jacobs control box

adjusted to ensure optimum performance, efficiency, and safety of the wind-electric system. It's a watchdog that does more than bark when something needs to be done. Furthermore, it liberates you from sitting there and doing these things yourself.

GOVERNOR

Not all windplant controls are located in the control box, however; two very important operations must still be attended to: automatic governing of the aeroturbine, and manual shutdown.

It might at first seem simple to deal with a high wind and fully charged batteries—just pull the switch at the control box and disconnect windplant from batteries. Well, it'd work for the batteries. It'd be death's drumroll for the aeroturbine, however! Why? Well, look at it this way: The wind's energy is extracted by the aeroturbine, and much of it goes to the generator, where it produces electrical power. Make sure you understand this. The wind is *not* just turning the windspinner and turning the generator; it takes power to produce power, horsepower for watts. Now, if the aeroturbine tries to extract all the wind's energy, what happens? The wind stops and, eventually, so does the aeroturbine, right? Well, the same holds true for the generator; if it tries to extract all of the energy from the aeroturbine, the aeroturbine will stop and, inevitably, so will the generator. We have to leave enough energy in the aeroturbine so that it continues to spin, just as we must leave enough energy in the wind so that it will move on. But we should all agree that the greatest amount of energy is going to the production of watts from horsepower in the generator.

We all know, just from what's been written thus far, that the battery is what stores the energy gath-

Fig. 3-24. A Jacobs governor. At high speeds, the spring-loaded centrifugal weights fly out, engaging the gears which turn the three blade shafts simultaneously.

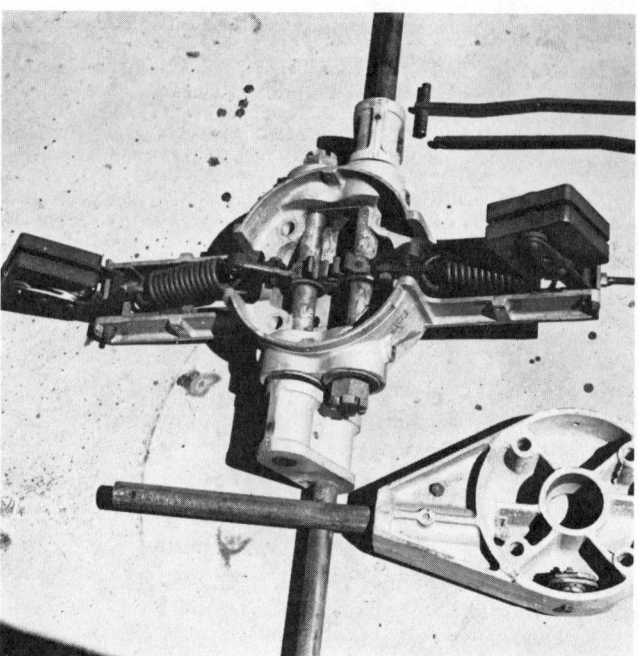

Fig. 3-25. A Wincharger centrifugally-activated blade-spoiling governor

ered from the wind—not as electricity, you'll recall, but as a chemical action. Nevertheless, the battery is taking energy from the generator. Now, what do you think is going to happen if we suddenly disconnect the battery from the generator? Does the generator have some way of storing the electricity it is producing? No, of course not. So, what happens? Well, it stops producing power (yes, it *will* still produce voltage, but *not* power), since it has nowhere to put it. Make sense? And, if it stops producing power, it must stop *using* power (which, in a manner of speaking, it does when it converts horsepower to watts). This is reflected back to the aeroturbine.

Unexpectedly, the aeroturbine has vast quantities of power that isn't going anywhere. So, in turn it gives the power back to the wind, right? Wrong! That wind has already moved on, leaving no forwarding address. The aeroturbine is stuck with it. What does it do with it? It makes itself spin still faster. Centrifugal forces strain fasteners, stressing and fatiguing the metal. Suddenly, what would take pages to describe as a sequence occurs in a split second. To a casual observer, the effect is no less dramatic than if someone had strapped a stick of dynamite to the whirling blades. Kkkrraaannnngggg! Splinters scythe holes in the dark sky. Ricochets scream in the night as a rain of metal impacts roofs, rocks, your neighbor's new Mercedes. There's the dull thud as bolts and wire bury themselves in the earth, the nearest trees, and your neighbor's pedigreed Siamese. And then silence, save for the gnashing of teeth. And all because you built a windplant that had no automatic blade governor!

You don't get out of the experimental stage until you have thoroughly accounted for rpm control of the aeroturbine. I almost finished that sentence with ". . . in high windspeeds," but, as we saw in the last section, "unloading" the system suddenly can seriously jeopardize the windplant even in windspeeds within the operational range. It doesn't matter whether it's a blown fuse or a gust of wind—without governing, the windplant dies just as readily. We can govern a windplant in one of five basic ways: spoiling, braking, sidefacing, stalling, and loading.

Spoiling is also called "feathering," a term for what is done to keep the prop on a dead engine on an airplane in flight from "windmilling." Applied to a windplant, this is a process that progressively interferes with the aeroturbine's efficiency in extracting energy from the wind. It occurs by changing the pitch of the blades (Fig. 3-25) or the angle they make respective to the plane in which they rotate; in high-tech lingo, they no longer lift, they drag. The desired effect—rpm limitation—is achieved.

Fig. 3-26. Attaching an airbrake governor to a 1250 watt Wincharger prior to raising

Braking is just that—some counterforce (usually friction) is applied, and the windplant slows. This can be a friction brake of the type used in automobiles. Another popular braking method is the airbrake; centrifugally activated vanes deploy and "scoop" air (Fig. 3-26), causing considerable drag (and sometimes noise too).

Side-facing is one of the more ancient and reliable means of limiting windplant rpm; almost all water-pumping wind machines employ this governor type (Fig. 3-27). Though there are two ways to do this, the more frequent technique is to "offset" the aeroturbine's axis of rotation from the lollyshaft (crystal-clear, right?). When the wind's speed reaches a predetermined value, the aeroturbine is forced (by wind pressure) to veer out of the wind; as the windspeed increases, the angle increases until, in high winds, the aeroturbine parallels the wind's direction and is virtually shut down. Another twist on this theme was incorporated into one of the windmachines of the pre-REA period. Instead of facing to the side, the aeroturbine tilted back and up (Fig. 3-28) in high winds; the propeller faced the falling rain!

Stalling is a technique demonstrated in many experimental windplants; another word for it, in that particular usage, is "laziness." Here, the windplant depends on the "stall" characteristics of the aeroturbine used to limit rpm to a safe value. With enough testing and experimentation, this is a possible governing method, but it works only with some types of aeroturbine and has a less sure effect compared

Fig. 3-27. Tail is in the side-faced governing position on this waterpumper.

with other governor methods.

Loading is sometimes described under the braking method of governing, because it is, in a manner of speaking, "electrical braking." If the windplant is developing power, this very occurrence is limiting windplant rpm, as previously described. Remove the load, and the windplant requires another means of governing. Knowing this, it is possible to electrically load a windplant, bringing on more and more of a load—whether more batteries of power-dissipating devices (such as heaters)—as the windspeed increases. This is tricky work, involving sensing methods and switching. A computer is better equipped to handle the decision-making required to protect and yet allow normal operation. Unfortunately, this still doesn't circumvent the possible circumstance of a blown fuse.

Brake, stall, and load governing share a fundamental fault. Their action is similar to braking an automobile with the gas pedal still depressed; rpm will be limited, but stresses are not. In high windspeeds, this means that windloading is not relieved.

Activating a governor at the right time is important. The most accurate means of doing this seems to be centrifugally (sensing rpm), and this technique

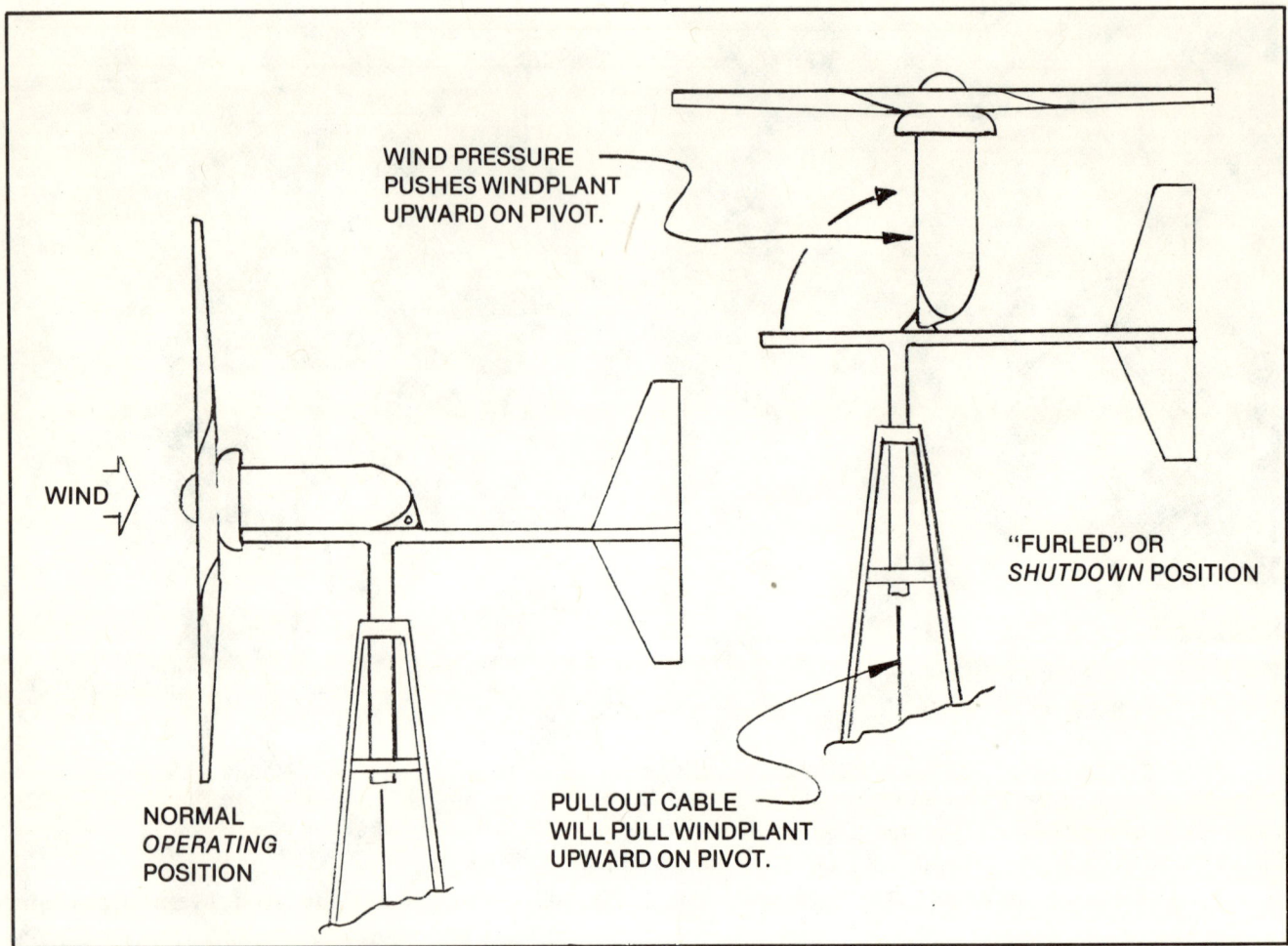

Fig. 3-28. "Up-facing" governor

is used extensively with the air-brake (braking) and pitch-change (spoiling) methods. Wind pressure is the cause of side-facing; the aeroturbine's cut-out point is determined for a given windspeed. Voltage, current, or windspeed may be sensed electrically to activate governors; this takes care of electrical braking (loading).

In my own experience, it's relatively easy to get a governor to activate; getting it to deactivate when the condition for its use is over causes the difficulty. The air-pressure (side-facing) technique has two problems along this line. First, since an unloaded aeroturbine can overspeed in windspeeds below its rated windspeed, a technique that's involved in sensing air pressure alone is not proper protection; if adjusted for this probability, it begins to side-face too soon. Second, it usually takes some fine engineering to overcome the fact that it takes less air pressure to hold the windplant side-faced than it does to actually side-face it initially; this means that it will not deactivate at the same point where it activated. Rpm-sensing, however, is usually quite responsive; the spring-loaded weights fly out when the centrifugal forces reach a certain strength (corresponding nicely with rpm) and, once the aeroturbine decreases in rpm below this point, the weights pull back in, deactivating the governor.

MANUAL SHUTDOWN

Every windplant needs a manual shutdown method that is operator-controlled. Whether you intend to do some work up there or wish to protect the windplant from a severe storm, this action must be positive and immediate. As well, you will use this system whenever the batteries are charged and there's no sense in running the windplant unnecessarily. When designing a windplant, some consideration might be given to combining the automatic governor with the manual shutdown method to avoid duplication. Whatever the technique, it should relieve windloading of the aeroturbine and tower; for this reason, side-facing is the most widely used manual shutdown. While it can be done electrically, a hand-cranked winch at the base of the tower connected by a wire with the mechanism on the windplant serves adequately. Check your late-night weather report for

WIND ENERGY 83

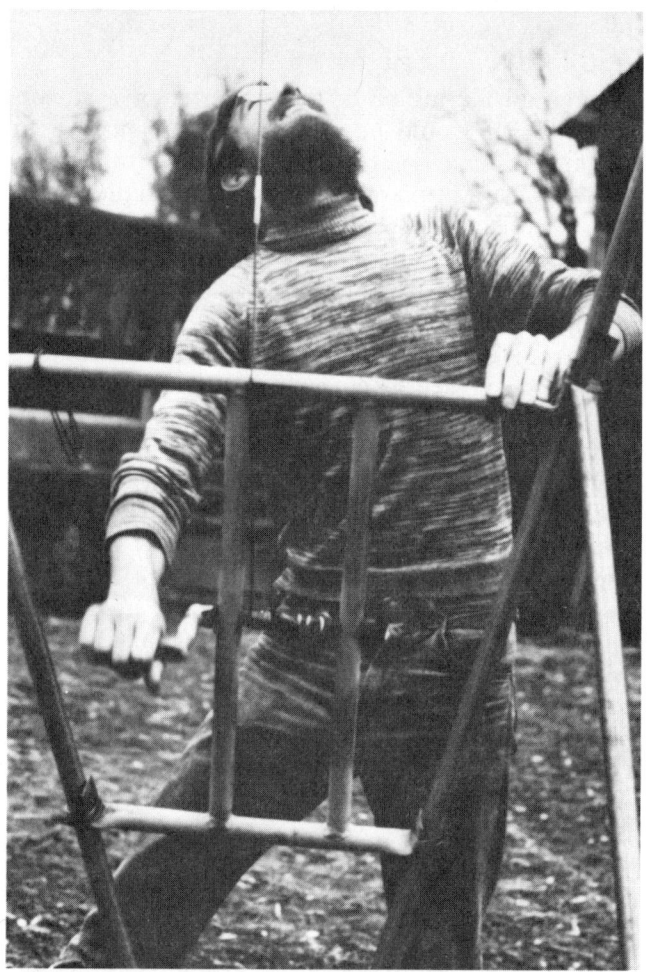

Fig. 3-29. The author using the manual shutdown winch

Fig. 3-30. After manual shutdown, the tail is parallel to the blades and orients them out of the wind.

storm warnings; "furling" the machine in a wild storm in the wee hours of the morning in your skimpy terrycloth robe is no fun.

USING WIND ELECTRICITY

Well, we've gone through gathering the wind's energy, generating electricity from it, transferring and storing it, and system and windplant control—now let's discuss using it! And the criteria that apply to wind-system design must not be forgotten here: we must *need* less energy or fork out a seemingly large sum of money to build a system which can compare (in quantity) with the level of consumption we can enjoy when we *rent* energy. Even our awareness that many of the costs in renting energy are hidden, while those of a wind-electric system are not, may not be of much consolation when we stare at the figures. As well, while we can learn to change our wasteful habits, we are nevertheless confronted with the fact that all of the machines we own (toaster, re-

frigerator, lights, etc.) still require 120 volts AC, 60-cycle. Or do they?

DIRECT USE

For certain, any device which directly uses 120 volts AC, 60-cycle, must use 120 volts, but it may not always require AC (alternating current), nor does it always require 60 cycles (120 changes of direction per second). Wherever electricity is changed into heat (toaster, oven, heaters, etc.) or light (incandescent), 120 volts DC (direct current) can be used; if you'll recall, this is what we can get from a battery directly. Light bulbs (incandescent) would last longer and burn brighter on DC, allowing use of lower wattage ratings for the same intensity of light; more lumens per watt means higher efficiency, too. Some motors that operate on 120 volts AC, 60 cycles, are "universal motors"; this means they were designed to operate on AC or DC. Stereos, record players, radios, etc. that are transistorized (solid-state) in many cases actually convert the 120 volts AC, 60 cycles, to a lower voltage (by using a transformer) and then rectify the AC into DC for use. For these devices, bypassing the transformer and diodes allows direct use of 12–24 volts DC. If you're definitely headed toward windpower, this might be a determining factor in selecting your next stereo system—its capacity to run off the voltage you select for the windplant!

INVERTERS

For the stuff that you can't use on DC or a lower voltage, an inverter is used. Inverters are electronic or electrical and they operate to change DC into 120 volts AC, 60 cycles, even from voltages lower than 120 volts DC. What happens is that the DC is changed to AC in a "vibrator" and, when fed into a transformer (which can only work with AC), it is stepped up to 120 volts. The vibrator can be made to operate on 400 cycles or 60 cycles, which means careful selection of surplus inverters for use in your system; operating a 60-cycle device from a 400-cycle inverter means minimum life and maximum smoke!

Inverters are expensive. The cheapest cost about a dollar for 3 watts, and the best over a dollar a watt. The larger the wattage, the more you tend to pay per watt delivered. Inverters are efficient only at their rated power; when operating light loads or idling (delivering no power), they consume a lot of energy, so extra wiring is necessary to switch them on or off. You must size the inverter to handle at least the heaviest wattage device that you'll operate from it; you'll have to size it even larger than this if you expect it to handle several things at one time. To avoid the gross inefficiencies (including wasted power) resulting from the operation of devices over a wide range of wattages, a smarter move is to use several inverters—a small one to handle wattages up to 200–500 and a larger one to handle the heaviest wattage you have. However it's done, it's obvious that converting DC to AC is an expensive venture—so, wherever possible, it should be avoided. It's interesting that inverting, say, 2,000 watts of DC to AC might run $2,000, where converting the equivalent power from AC to DC would only cost $2–3 at the most; the inverter would weigh probably 40 pounds and the "converter" (diode) a few ounces!

With sufficient forethought, it is possible to eliminate the need for anything but a very small inverter, if even that. This mostly depends on how much you have invested in what appliances, tools, etc., you already have. If a wind system would be hard pressed to power them anyway and you use them infrequently, the system's standby generator might be designed to handle them alone. For larger power-consuming devices, I'd suggest you try the alternatives before you try to put them on the poor old wind system! So that you have an idea of what I mean by alternatives, check out the rest of this book.

ASSESSING NEEDS

Before considering buying an inverter or a standby generator, it would be wise to assess your present power consumption. This involves determining the power consumption of every appliance, light, motor, etc., used in the home or shop; and determining the approximate amount of time each item is used per day, week, or month.

Just looking at your electric bill is one way that you can get a close approximation of the amount of power you presently consume; this is measured in kwh. "Kwh" is difficult to pronounce, so we say the equivalent—kilowatt hours. Akin to ah (ampere-hour), this describes how many kilowatts you've used for how many hours. By dividing the total for the month by 30 days, you'll know how much you use per day; a 300-kwh monthly consumption reduces to an average of 10 kwh per day. Further reduction is meaningless; you might have consumed 10 kilowatts in one hour or 1 kilowatt in ten hours. Since different appliances of varying wattage ratings are being used on and off during the day, we need a quicker way to figure how the kwh rating breaks down. Using the utility bill as an expression of your needs is not really a good way to start out planning a wind-electric system; this means that you are (mistakenly) equating your needs with your consumption. Everyone should realize that consumption can *always* be reduced without sacrificing necessities or comfort. Using wind means using less power, unless you have tons of money or tons of wind.

It makes sense to know what each of your appliances, lights, tools, or motors, really consumes over a month's time; it's the first step toward cutting back on real offenders. To do this, go around your house and get the wattage rating from each electric appliance—anything that uses electricity. Wattage or amperage, by law, must be stated somewhere on the device. If it gives you amperage, you'll read something like 7.7A or 2.3A; once you've multiplied this times the voltage, which is printed 110V or 117V or 120V, you have the wattage; if the device says 350W or 1340W, this is wattage (watts) and you do *not* multiply it times the voltage, because they've already done that for you. Sometimes, appliances use a term like VA instead of W for wattage; this stands for volt-amps and it means the same thing.

With a completed list of wattages for everything that consumes energy, we must now sit down and, being completely honest, figure out how much we use each device in a month's time. If we used it the same amount of time each day, we'd figure that out, and multiply it times 30 days to get our monthly figure. Some things, however, we use more on one day than we do another. More often, we use things consistently through the weekdays but change our use over the weekend. If you have a steady job, lightbulbs might see less use during the week (because you go to bed early) than on a weekend.

Let's look at some examples. You've already determined that your stereo consumes 120 watts; you always turn it on when you get home and listen to it while you sip a cold carrot juice, right? Then it goes off when you go out to the shop to work on a few projects. That figures out to 30 minutes of use weekdays; 30 minutes times 5 days per week times 4 weeks per month is 2.5 hours per week and 10 hours per month. How about weekends? Well, maybe 2 hours average per weekend; it varies so don't try to pin it down exactly. All right, that's 2 hours per weekend at 4 weekends, or 8 hours; combined with the weekday total per month, you have 18 hours of operation per month, approximately. At 120 watts, that's (120 multiplied by 18) 2,168 watt-hours per month.

How about the toaster? That's easy to figure, right? You have two pieces of toast every morning, your mate has two, each of the two children have one, and, you remember, you grownups share a piece at night; that's four times per day you operate the toaster, weekends and weekdays alike. Since the toaster is only operating for a few minutes, you'll need to time it for each cycle, so you do; it comes to 2.5 minutes for your particular toaster. So, at four times per day, it's on 10 minutes per day. Times 30 days, you get 300 minutes per month or, dividing by 60 minutes for each hour, 5 hours per month that you operate the toaster. We've already discovered that the toaster consumes a whopping 1,200 watts, so that's 5 × 1,200 = 6,000 watt-hours per month.

Figures in watt-hours aren't much use to us; we must convert them to kilowatt-hours. Since one kilowatt-hour equals 1,000 watt-hours, we simply divide our watt-hour figures by 1,000. Oh, darn, more division. But remember that dividing by 1,000 means you only have to move the decimal point three spaces to the left. So, our 2,168 watt-hours of stereo pleasure per month equals 2.168 kilowatt-hours or, since it's so close, we'll round off to the nearest figure—we use the stereo 2.2 kwh per month. The toaster, at 6,000 watt-hours, figures to 6.0 kwh per month. When you've got them all figured, add them together and compare that figure with your present bill to see how close you've come to figuring out how much energy you consume.

CUTTING DOWN

Although it is difficult to generalize about the amount of power a windplant will develop in a specific month's time, a ballpark figure may be reached and presented in kwh. A comparison of your present consumption vs. the power production of even a fairly large windplant (2.5 kilowatt range) will probably indicate a gap that will leave you gasping. For those of you simply exploring the fantasy of owning and operating a windplant, this will inevitably cause you to insert all of the material back into a very big envelope and store it away for future (like in 2,000 years) use. For the borderline cases—"maybe I'll do it and maybe I won't"—it's a matter of looking at each side. On the one hand is the possibility of enlarging the aeroturbine, putting it on a higher tower, using a more efficient type of aeroturbine, or putting up several aeroturbines. The cost here usually gets you to check out the other hand—cutting back on the energy consumption for both home and individual use.

Most folks who consume power at present-day levels feel that anything less than, perhaps, a 5% cutback will result in extreme personal sacrifice. In many cases, this is not true; a much more reasonable estimate would be that cutbacks on the order of at least 50% can occur with little or no discomfort or inconvenience. This comes about by sufficient knowledge of how we waste (through our own habits), how the appliances we use waste, and how we are wasteful in our misuse of these appliances. When you start thinking you've done all you can to stop the waste, it's time to do more reading.

Whole books have been written on the ways in which we waste or consume energy; I refer you to the Sources and References section for examples. Heating in winter and air conditioning in summer are two big consumers of energy, and they need not be. The money spent on the equipment and the fuel (electricity, coal, oil, etc.) would be better spent on insulation or natural cooling and heating techniques (in home construction, airflow, etc.). If you think that folks a long time ago didn't enjoy summer coolness and winter warmth, you are wrong. The kind of information required to do it as they did isn't lost, but it's only now becoming readily available. For all of the controversy in the back-to-the-land movement, there can be no doubt that it has revived the spirit of frugal living and the common sense and skills associated with it.

AUXILIARY POWER

It would be unnecessarily expensive to build a wind system that would provide enough electrical power to meet all our needs, even in the calmest months of the year. For this reason, if no other, a standby generator is a must. With it, we can take care of the "peak" loads in the doldrums with a much smaller and less expensive windplant. To some this might seem a decadent luxury, but the standby generator unit (hereafter SGU) provides many more necessary services. Batteries should be periodically charged to capacity; called the "equalizing charge," this assures their longevity and prevents sulfation, a condition which affects a battery that has been improperly charged. Even in the windiest climates, a relatively calm period of the year occurs, and it would tax even the largest windplant to perform the equalizing charge.

The SGU is a backup system for the windplant; if you need to service the wind-electric system, the SGU maintains the battery bank's charge. During emergencies, where additional power is required, the SGU really earns its keep. Or, it may be that the SGU is able to deliver a different type of power as required; rarely used tools that operate at 110 volts AC, 60 cycles, may receive electricity from the SGU even though the wind-electric system and all household uses are 32 volts.

Sizing the SGU is a matter requiring your complete attention; its rating must be adequate for the largest tool you may need to operate independently of the wind system. If it functions only as a battery charger, it need not be large at all; in most cases, the SGU has a smaller rating than the windplant. If you can muffle the engine noise or isolate the SGU some distance from the house or workplace, a small SGU running a long time will prove just as adequate as a larger one running a short period of time.

So, what does an SGU look like? Usually, we have a generator or alternator coupled, directly or through a gear/V-belt/chain drive, to a small engine fueled by propane, methane, steam, diesel fuel, or gasoline, with the latter two being the most commonly used. If it's a commercially built unit, you figure what you want or need, and buy it. There is an alternative if your SGU need is less than 1,500 watts —a homebuilt version. This involves purchasing a 5-horsepower (horizontal axis, four-cycle) engine and selecting a good car alternator. Together with voltage regulator, ammeter, switches, and a V-belt, this is bolted to a wood or welded metal stand. You have to do a good job to equal the performance and reliability of the manufactured assembly, but it's well worth it; even standby generators don't work sometimes, and a homebuilt unit is somehow easier to understand, maintain, and repair than its store-bought counterpart.

If you're installing the windplant in a remote area, the first item you purchase/build and set up may well be the SGU; it will help you set up the windplant and keep you and your batteries supplied with power in the meanwhile.

TOWERS

You can't see the wind—only its effect on things—but, for a moment, let's pretend that you can. What do you see? Probably the first thing that you'd notice is that it moves more slowly at ground level. Know why? Even if the ground is level and fairly smooth, it will provide friction (or a resisting surface) to the free agent of moving air, and therefore slow it down. This slow-moving air will, in turn, provide friction for the air above it, and slow *it* down. This goes on and on. If the wind were constant in both direction and speed, we might be able to see barely distinguishable boundaries between the layers of faster and slower moving air. Normal wind will, however, fluctuate in speed and direction, so these boundaries will be constantly changing, and this changing has a name: turbulence. If you were standing in the midst of it, you would not fully appreciate its effect; in most instances, it isn't exactly going to sock you around or throw you to the ground. To an aeroturbine, however, any disruption of a flow of air will steal power it might otherwise use. With some types of aeroturbine, turbulence is more than power-robbing; it can cause some potentially destructive stresses on

Fig. 3-31. A horizontal-axis engine suitable for use in a homebuilt standby generator

Fig. 3-32. A heavy-duty military surplus standby generator

the machine's components.

Turbulence will be more pronounced with rougher terrain, the more obstructions there are above the ground, the higher the windspeed, and the more gusty it might be, so it's not something to be ignored or considered insignificant. Rudimentary observations will show detectable differences in windspeed at even 10-foot intervals of height above ground. A formula has been developed to aid in figuring out what the windspeed would be, say, at 45 feet above the ground, for a given windspeed at ground level (see Cubbyhole, Section F); this applies to a flat, level plain, but it can be somewhat of an indicator in obstruction-ridden or rough ground conditions, too. In the wind-energy field, 30 feet of tower is considered to be the minimum required, and few exceed 80 feet in height.

Some folks shy away from towers, certain that they could not or would not climb one. This is an understandable feeling. However, this should not influence whether you use a tower or not; there are always some crazy people around who like nothing better than scrambling up towers to get a new perspective on things. Besides, with a correctly installed machine, only *annual* trips up the tower are required. This is no more involved than changing gearbox oil (if your windplant has a gearbox), shooting the zerk fittings with some grease (if the aeroturbine has these), and a casual check for loose bolts, screws, wires, etc. Every three to five years, the blades and/or windplant may need a coat of paint. So, the initial installation is the only time you really need to hassle with the tower in any way; there are people who raise towers and windmachines both, and they can bring the proper equipment and generally take the worry out of it for you.

FIGURING TOWER HEIGHT AND SITE

There's a bit of interplay between tower height and siting. Finding and buying a tower means that you're stuck with whatever you find, but at least you'll know whether it's too tall or short for your situation. If you're building your own tower, the following criteria will give you the ballpark height to aim for:

1. The aeroturbine, and its tower, should be located as close to the batteries as possible; this cuts down on the need for a lot of electric wire and resul-

Fig. 3-33. Assembling a homebuilt octahedron module tower

Fig. 3-34. A sturdy ownerbuilt wooden tower for a waterpumper

tant I^2R losses (power that is robbed by the resistance of the electrical wire). For the same reason, the batteries should be located as close as possible to the motors, lights, appliances, etc. that use this power. With long runs of electrical wire, the only way to decrease the losses is to go to a larger size of wire, and this will cost more.

2. The tower should rise 15 feet above all obstructions within 400 feet. This is a tough one to get around, and if it's just not possible for your situation, then give strong consideration to placing the tower where the aeroturbine will at least have unobstructed access to the *prevailing* winds in your area. Climatological stations usually keep a record of wind directions throughout the year; this information is available to the public, so consult it before installing any windpower system. Be aware, however, that this information results from a few minutes of reading a few times each day (for most stations) and depends on how high the instruments are mounted, what kind of instruments are used, how long it's been since they were calibrated, and where they're located on the property. The information is better than nothing, but how much does that tell you? Additionally, it's a well-known fact that substantially different readings of windspeed and direction can be obtained at a particular site at points that may be only 50–100 feet apart.

3. A higher tower is going to cost more money. A lot depends on the type of tower used, but be aware that it's not a linear cost per foot of tower height. My own guestimation seems to indicate that the price goes up with the multiple of the increase factor; i.e., a 40-foot tower will cost four times as much as a 20-footer, and a 60-footer will cost six times as much as a 20-footer. We're not just dealing with the tower materials in this formula. A free-standing tower (legs cemented in the earth) must be spread out more at the base with increased height, and a guyed tower requires more and longer lengths of guy wire. This is not intended as a rule of thumb, and it certainly isn't a direct proportion, so check it out. If your pocketbook isn't too tight, go for as high as tower as you can find and know you can install; if you're going to be the one climbing up and down it, get one that suits how high you want to go.

4. Towers are either built from the ground up or raised fully assembled into place, sometimes with the windmachine attached. Only the beefy three- and four-leg steel or wood towers can be built from the ground up or, if raised assembled, support a windmachine attached to the top during the raising. It's very embarrassing to buy or make a tower and then find that there is no way you can stretch it out and raise it, too; check out the site for maneuvering room. For a guyed tower, you must eyeball all the places

Fig. 3–35. An enthusiastic tower-raising crew, pre-raise.

where the ground anchors will be installed and guys attached; getting those five equally spaced spots can be tougher than it looks.

It is highly recommended that you *not* raise a tower or windmachine without first reading (and studying) procedures and information that has been written on the subject (see Sources and References section). It's not particularly complex, but it's very easy to get into a dangerous situation without knowing it.

NOTES TO DIYERS

A few words of advice for the homebrew-oriented individuals: stay realistic. A little bit of knowledge is dangerous. If it doesn't cost you a wind machine outright, some other stuff in the local vicinity of a windplant, or a life, it can lose you a lot of personal energy and put you on an anti-windpower trip. I'd rather see most people try to build their own windmachine because I feel that it can be of great benefit to them, as well as a lot of fun. But I've recommended buying one to some people I've met. Sure, anybody can learn something. I can be an astronaut and you a president. The only problem with that notion is that

WIND ENERGY

Fig. 3-36. Raising a 400-lb Jacobs wind-electric generator using a machine-raising gin pole.

Fig. 3-37. Gearing, jackshaft, and alternator arrangement on Earthmind's S-rotor

it's just that—the way it *can* be, which is not the same as the way it *is*. I've met some folks who'd kill themselves if they tried it. So I don't encourage them. I'm not worried about the ones who prove me wrong. Just the ones who prove me right!

Plans for windplants abound, and enough good designs exist to give you ideas. But if you're prototyping your own windplant, expose it to the hard analysis of others around you, especially those working the wind-energy field. Don't be afraid that someone is going to steal your idea and patent it! I've had many folks contact me who after only working in the field for a year or so have found the "breakthrough of the century." Sure, it happens, but not for any of the ones I've seen; correctly applied criticism will refine any good idea and will, except for the very stubborn, junk out any wild ideas that will simply not work. This will save you time, energy, and the least expendable thing—money!

About hardware. We'd all like to dig in there and wire our own generator, cast our own generator housing, manufacture our own bearings, and so on. This is silly! If you're trying to figure out what to do with the rest of your life, okay; otherwise, look around for what's available off the shelf or adaptable. Use everything you know to help the design; don't do things the hardest way, just the most sensible way. Special items cost more, whether you buy them or have them made; avoid this if possible because money evaporates on big projects fast enough. On the other hand, don't get psyched with prices; if you really need it, scrape it together and buy it. Money isn't the only criterion; time, skills, tools, and the amount of energy it takes you to do it another way should all be factors in the decision.

WIND FOR PUMPING WATER

The real success stories in windmachines today are the units designed to pump water. Pumping water with the wind is much easier than generating electricity. For verification, note the number of water-pumping windmachines you've seen recently. If you can't count more than three or four, you must be in downtown L.A. Although a misnomer, the "farm windmill" is the name given to this turbine. They're found nestled into the trees near an old farmhouse, or sitting alone in the midst of nowhere with a huge cattle-watering tank nearby—which is tribute to their reliability, since you don't isolate something which is finicky or temperamental. While occasionally equipped to pump from a shallow well or stream, the typical water pumper is lifting water from a steep well (75–250 feet), and the largest units can pump water from more than 1,000 feet down!

The basic water-pumping setup includes three subgroups—above-ground, ground-level, and in-well (see Fig. 3-38). The above-ground portion is the tower and windmachine. The twelve-to-twenty-four-bladed turbine is attached to a gearbox, where the rotary motion is converted into a reciprocating (up-and-

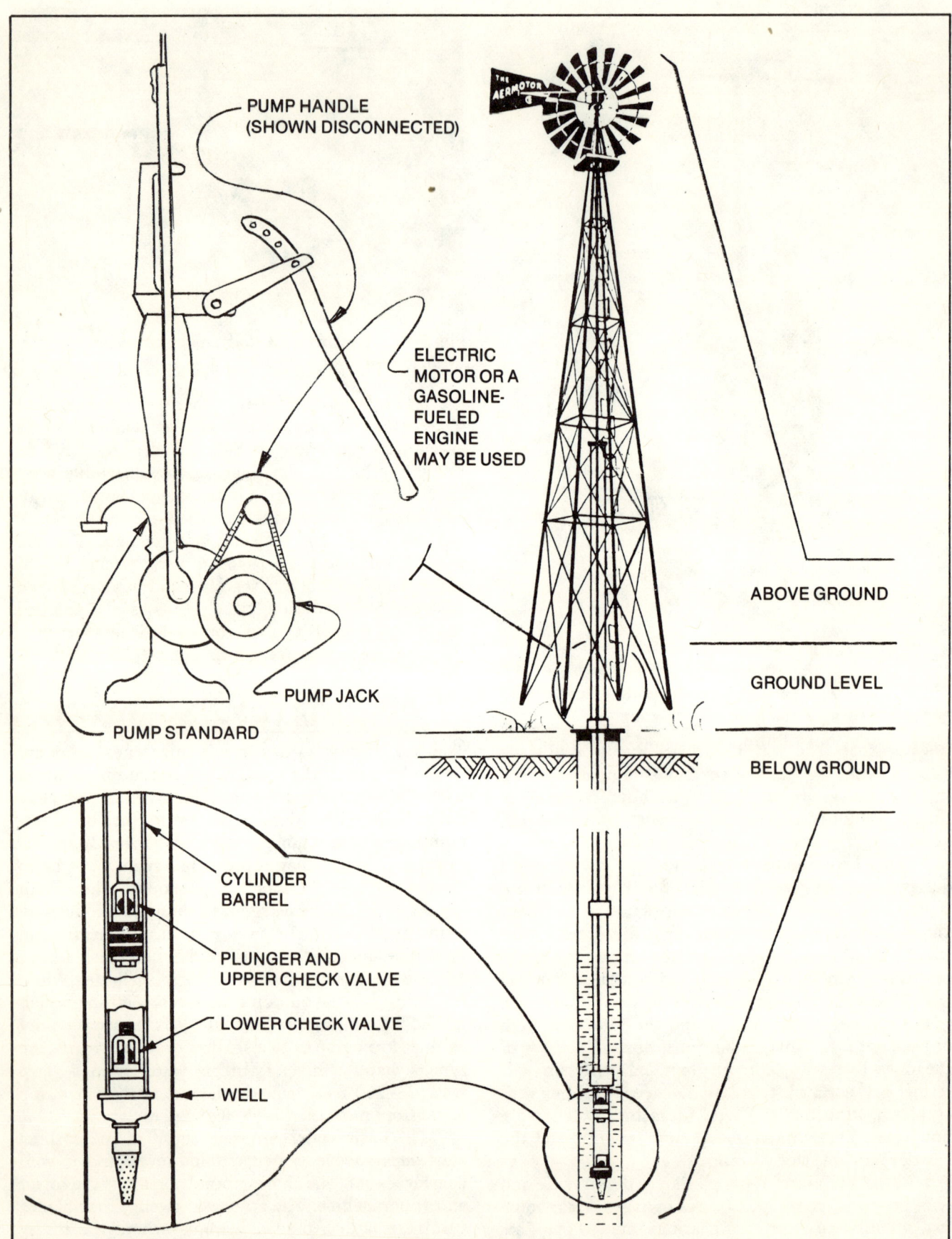

Fig. 3-38. A wind-powered water-pumping system

down) motion. A mechanical linkage—wood or metal sucker rod—connects the aerotrubine with over-the-well equipment at the tower's base. At ground level, we have the hole in the ground which is the well and a pump standard or stuffing box. The pump standard is the old-fashioned hand-pumping verison of the stuffing box; attach the handle and, if there's no wind, you can still get water out of the well. The pump standard serves several other functions. Water can be directed out of a spout or into buckets for dousing heads on hot days, or through a pipe to a water-storage tank, for use when there's no wind and your arm can't handle the demand. As well, the pump standard can be thought of as an anchor, suspending the in-well hardware off the well's bottom. This means you can retrieve the assembly, should you have to do so, and it keeps sediment out of the pump workings.

The in-well machinery is composed of the water pipe, up through which the water is moved. The foot valve, pump mechanism, and pump rod are designed to lower through the pump standard and through the water pipe; this facilitates easy removal for replacing the wearable parts of the system. Once inserted back into the well, the foot valve serves as a lower check valve; combined with the action of the check valve in the pump mechanism itself, it will work as a true pump. The up-and-down motion of the sucker rod (aeroturbine to pump standard), the pump rod (pump standard to pump mechanism), and the pump mechanism (riding like a piston in the cylinder) gives us the necessary action. With the upstroke, water is brought into the cylinder; the downstroke pushes this through the upper check valve. The next upstroke brings in more water and, at the same time, lifts the water in the pipe above the upper valve. Again and again it repeats until the water is coming out the spout or on its way to the tank.

Water-pumping windmachines are a good investment, even in low-wind areas. They're rugged, with little to go wrong. Water pumpers are not plagued with the change-it-completely-once-a-year attitude held by automobile manufacturers. With no major change in the past fifty years, parts are neither hard to get nor grossly expensive. Tower height and aeroturbine size are determined by the depth of well, the height of surrounding trees, and the amount of wind the area experiences; freestanding metal towers or homebuilt wood towers will readily accept the machine.

Alternate pumping methods are available once the basic installation—ground-level and in-well—has been completed. A pump jack can be attached to the pump standard and stuffing box alike to provide the necessary reciprocating motion from rotary motion. Then, it remains for you to attach a small gasoline-

Fig. 3-39. An early waterpumper in the Midwest

or propane-fueled engine or an electric motor to drive the pump jack. For a little more money, this system beats all other owner-installed water systems. Although the storage tank is an added cost, it allows for gravity-flow water pressure to further save energy; in any rural area, this is a good backup water source in case of a grass, brush, or forest fire (when power lines are the *first* things to burn). A 20-gallon-a-minute well is useless if the submersible pump has no utility electricity to power it. On the other hand, firefighters and homeowners alike can pull from a storage tank to fight a fire.

The windsail and S-rotor are good alternatives to the manufactured water pumpers for folks willing and able to build their own systems. Deep-well systems will need reciprocating motion, so don't forget that conversion from rotary motion. If your best wind site is not your best well site, don't despair. Offset well systems using wire transmissions and counterweights have been used for years, with good success up to a quarter-mile. Further refinements, like disconnecting the aeroturbine from the well pump when the tank is full, have been designed and used, with emphasis on nonelectric sensors and controls.

	System One	System Two	System Three
Windplant	2500 watt, 110 volt D.C. 3-blade, Jacobs (restored). Rated windspeed – 25 mph. In 8 mph AAW, will produce 300 KWH per month (max.) *$2000*	1200 watt, 32 volt D.C. 2-blade Wincharger, (restored). Rated windspeed – 23 mph. In 8 mph AAW, will produce 140 KWH per month (max.) *$1000*	200 watt, 12 volt D.C. 2-blade Winco (new). Rated windspeed – 23 mph. In 8 mph AAW, will produce 45 KHW per month (max.) *$500*
Tower	50-foot, 4 leg • free standing • concrete base • manufactured *$1000*	50-foot octahedron module tower • free standing • concrete base • metal tube • homebuilt *$200*	40-foot, single-leg, wood, homebuilt *$90*
Batteries	19-6 volt, 220 AH Deep cycle *$1000*	6-6 volt, 220 AH Deep cycle *$300*	2-6 volt, 220 AH Deep cycle *$100*
Standby Generators	1500 watt, 110 VAC manufactured *$300*	600 watt, 32 VDC Homebuilt *$150*	400-watt, 12 VDC Homebuilt *$100*
Assorted Hardware	Concrete, electrical wire, lightning arrestors, etc. *$500*	Concrete, electrical wire, lightning arrestors, etc. *$300*	Electrical wiring, lightning arrestors, etc. *$100*
Total	$4,800	$1,950	$890

Fig. 3-40. The cost of three wind-electric systems in 1979

There are many advantages, as you can see, to wind-powered water pumpers beside the poetry-in-motion of the windmachine itself. If you've got a decision ahead about your own water system, it'd be well worth your time to further check into this kind of system. And, if you can get past the initial cost, it'll pay for itself in the long run with reliability, low maintenance, and good looks.

WINDPLANT COSTS

Okay, by now you must want some hard dollars-and-cents info. I've looked over a number of ads and brochures and checked with a few friends, and, along with my own contacts and sources, it reduces to the data given here. This will vary according to how deep you scrape to get the best deal, and how much of this information you've assimilated to help weed out the good from the mediocre. As of mid-1979, I've computed the cost for three systems (see Fig. 3-40), starting with one of the largest windplants available and proceeding to the frugal version of wind-energy usage. Along with everything else, except love and knowledge, these prices will go nowhere but up in the future. Don't put off your decision too long.

Each windplant is shown with a different tower, assorted (support) hardware, and a unique standby generator. Moreover, some of the stuff may be homebuilt or store-bought. So that you'll understand each of the footnotes, they're listed here:

1. All windplants are rated for maximum power output at 22 mph windspeed.

Fig. 3-41. A medium-cost, medium-power Wincharger (1200 watt, 32 volt) before being raised on the octahedron module tower in background

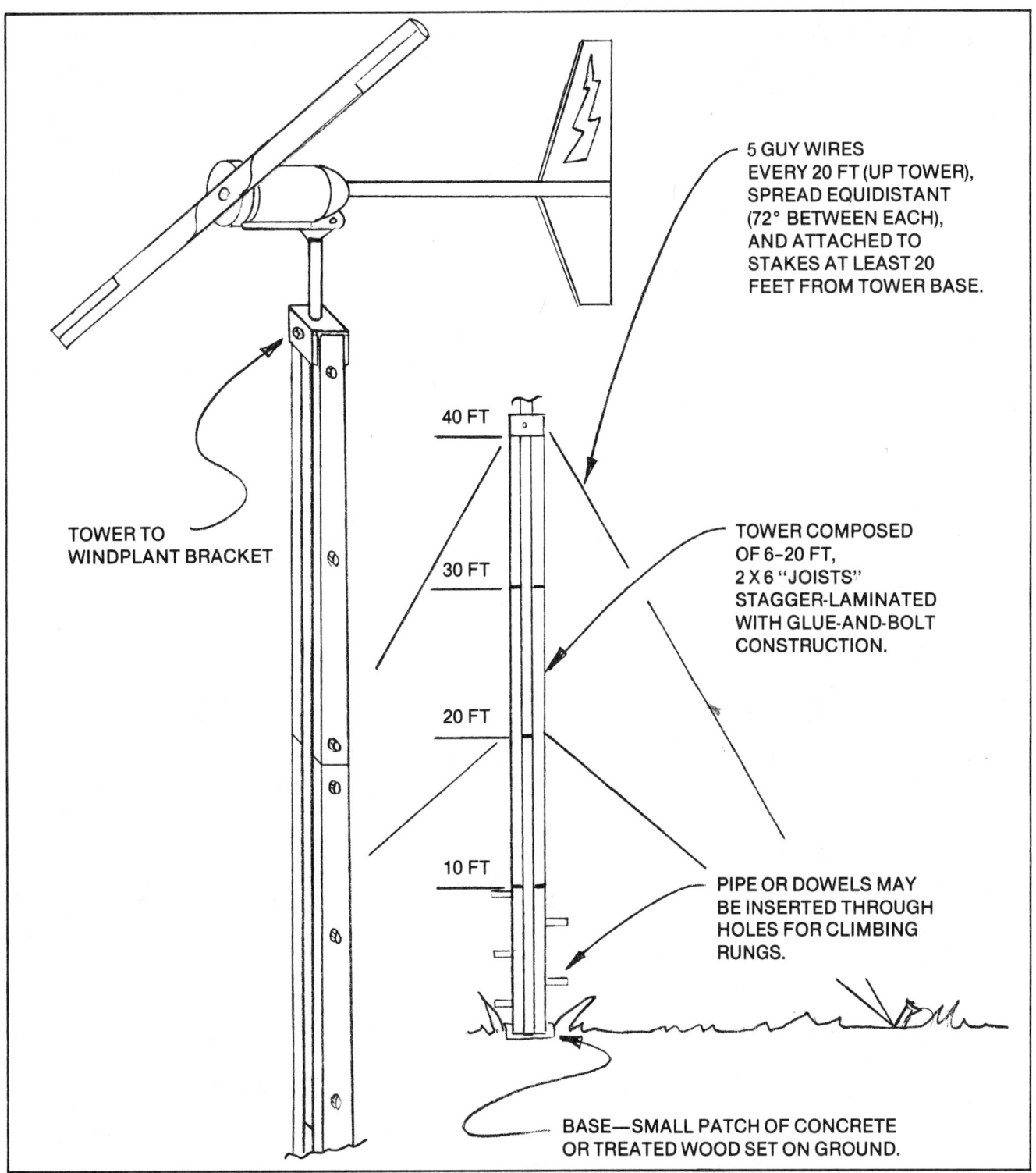

Fig. 3-42. A 12 volt, 200 watt windplant on a homebuilt, single-leg, guyed wood tower

2. The largest windplant uses a manufactured four-leg tower, the medium-sized windplant sits on a homebuilt octahedron-module tower (Fig. 3-41, and see Sources and References section), and the small unit is a homebuilt (guyed) wood tower (Fig. 3-42).

3. Batteries are deep-cycle, plastic-cased (not clear), twenty-hour rating.

4. Assorted hardware (if applicable) includes, but is not limited to, concrete for the tower-leg footings, guy wires, electrical wire (windplant to batteries), lightning arrestors, bolts and nuts, primer, paint,

rust inhibitor, grease, battery connectors, anemometer (windspeed indicator), etc.

5. Windplant includes aeroturbine, automatic governor, generator, slipring assembly, lollyshaft, tail (if used), pullout crank and wire, and control box.

6. SGU (standby generator unit) includes engine and generator/alternator (store-bought or homemade), voltmeter, ammeter, controls, fuel tank, and electrical wire to control box (or batteries).

FINAL NOTES

A full commitment to windpower is not a requirement for using it; in and of itself, it cannot be expected to supply all your power, except in a very few areas. In some states, there's a tax credit (as much as 55%) for a windpower installation as well as for solar-energy hardware; check it out if it makes a difference to you. Buy enough equipment to get started and see if it fits you. If it does, you'll know a little more about what more you need and, in joining the windfreak society, you'll get a better deal next time around. If it doesn't fit, just enjoy what you have as a conversation piece, or sell it; there are enough crazy people around who will buy it. And, unlike the horrible depreciation a car leaving the showroom floor experiences, windplants maintain or increase their value with age; it might be as good an investment as land. Happy hunting and, once involved, may all your days be windy ones.

4. Wood, Water, and Methane

Solar and wind energy are creatures of the air, coming to us on "wings." The three energy sources discussed in this chapter—wood, water, and methane—are earthbound, which we'd tap or harvest to use. Just because they're grouped in one chapter doesn't mean they're less abundant, practical, or accessible than solar or wind utilization. All depends on where you live!

WOOD ENERGY

Using wood as a source of fuel for heat and light is the story of humanity. It has only been a second in galactic time since we puzzled at the flames which licked the lightning-struck tree, and howled with pain trying to pluck them away with our fingers. With this indoctrination we found something which pushed back the night, giving warmth, security, and (eventually) a means of making meat more palatable. Little wonder we feel a primitive pleasure when we sit by the campfire or before the fireplace; it's part of our heritage.

Wood has been used so extensively throughout time, that it's a wonder we have any left. Part of the explanation for its persistence is the large yield of energy that even small quantities deliver. Another is that wood is a renewable resource, able to propagate itself without much help from (and despite) human beings; admittedly, it can be burned much faster than it grows, but the world is yet large. Because it is, unlike the sun, a finite resource, it must be treated as just that; wood-growth management is essential to centuries-long use. Unfortunately, at the present rate of use, wood will become a diminishing resource in the not too distant future. This is due to its popularity as a raw material for industry. Whether for tract houses or a forest of books, the use is high and the waste even higher.

For this reason, it is my contention that wood should be used only as secondary or backup source of heat to the alternatives of wind and solar usage. True, a complete wood-energy-based system would cost less money to install than a wind or solar system. And if none of the other alternatives explored in this book were available, this would be acceptable. In comparison to the utility-scale use of oil, gas, coal, and nuclear energy, wood use would be downright ecological. But with widespread use, wood supplies would be depleted and you'd only be robbing your children or grandchildren of a good backup source of heat in the future. If you tend to be myopic, consider the effort you must expend year after year to maintain the wood supply for your ovens, stoves, fireplaces, etc.

Wood heat, on the other hand, is an excellent secondary system to solar heat. Having trees about the place is not what I mean. If you've planned ahead and have a good stack of wood, it can bring you through the peak cold periods of the year and supply emergency heating as needed. It's fast and requires no special effort to start or maintain. And, as we shall discuss in this chapter, it can be readily adapted to space and water heating and to cooking. A hot meal, a hot bath or shower, and a warm habitat will turn just about any kind of emergency situation into a welcome adventure.

WOOD AVAILABILITY

There's not much point in talking about how to use wood heat or how much it yields if it's not available to you. If it's not, you might as well go on to the next alternative source. Basically, there are three ways of getting wood: buying it, scrounging someone else's, or growing and using your own. Let's check them out.

Buying wood. There are people who cut wood for sale to people who can't, or won't, or don't have access to it. Many times, this is a local project—the wood is cut from the seller's own land, a national forest, or someone else's land. Most of these folks do it as a living in their own area because the wood is there and there is a demand from some people in the community; access and good sense don't always motivate people to go out and get it themselves. If there's a city within a few hundred miles, it usually pays to go there to peddle the cut wood, since higher prices can be asked of people who have fireplaces and like to use them once in a while. Going so far is a lot of work, however, and with the increasing cost of gasoline there'll be less and less of this.

Some companies have noted the high prices city folk will pay for wood and they've introduced substitutes which will burn in the fireplace. Sometimes this is just newspaper that's tightly rolled and saturated with old oil; this is the low-grade stuff. Logs made from pressed sawdust and wax make good use of waste materials from lumber mills and will produce higher yields of energy than even good hardwoods, but they're expensive, cannot be cooked over, and cannot satisfy a larger market for the product.

Buying wood is not recommended unless you know where it came from, or you might be inadvertently supporting some of the horrible lumbering processes that occur today. If you're scared off from cutting your own because of the equipment cost, check it out; you might be surprised how quickly the self-cut wood will pay for the expenditure on the necessary tools at the rate you're paying someone else to pay off their investment *and* make a living. There's an old saying: "Wood warms you twice—once when you cut it, and again when you burn it." Besides the responsibility you assume for your own energy, it can be fun and certainly healthy—providing exercise, being outside, balancing physical work with mental. If you must buy, shop around and learn the source of the wood you're getting.

Scrounging wood. If you get a permit from the Forest Service to take wood from a national forest, this wouldn't really be scrounging wood, but it's the only word I could come up with that reasonably fits all of the methods I want to discuss here. If you do it this way—with a forestry permit—it's supposed to be for private use only; meaning no resale. In most cases, the wood is down and you just find it, cut it up, and haul it out of there. A certain number of trees always fall over of their own accord, but be prepared for a shock if where you're going is a recently lumbered area. It used to be that trees were marked and cut and dragged out, or rolled down slopes. I guess that took care of the easy-to-reach areas, and, with that excuse, everything nowadays is slash-and-burn. Why pull trees between trees that aren't useful for milling? Cut down everything, pull out what you want, and leave the rest or burn it. You might find some consolation in that they leave it for you or whoever comes along, but I didn't and don't. Nevertheless, this is where you might be sent. If you see some nice poles that you'd really hate to cut up, check back with the Forest Service to see if you can remove them as poles. You'll pay for it, but it's a pittance; I think we paid $12 for seventy-five poles last year, all over 12 feet in length and 6–8 inches in diameter. Best bargain since nickel yogurt. Don't attempt to remove poles with only a firewood permit; if you're caught, the penalty is severe and you'd deserve it.

All forestry wood is not downed; whether it's poles or firewood you're after, you may have to fell it. In 99% of the cases, you'll be thinning a dense stand of trees. In our area, the rangers mark the ones they don't want with blue paint, and I think they're generous in quantity and quality; there's absolutely no reason to cut unmarked ones. There's a technique to cutting downed wood, and about ten times more technique to cutting vertical stuff. Know what you're doing! Usually you're going to have a crew with you (or at least one other person, unless you're a fool), and for their sake if not your own, you should know what can happen and how, and what you're going to do to make that tree go where and do what you say. No place for amateurs. There are old woodcutters and there are bold woodcutters but there are no old, bold woodcutters.

Our forestry permit gives us permission to cut 8 cords of firewood a year for free, and we renew that permit each spring. If you're using a chain saw, it must have an approved spark arrester on it. Furthermore, you have to bring it in to get a sticker that says it's been inspected for compliance (also renewed annually). There are only certain times you can cut, you must pile the slash if you're cutting down trees, and a shovel and ax must be handy to your work area (in case of fire). You can only cut in approved areas, too. Some folks think that's too many rules, but I think it's absurdly minimal for the privilege of acquiring (in 8 cords) 100–200 million Btu's of heat energy. What do you think?

Not everyone is going to be fortunate enough to have a national forest close enough from which to get wood. So look for man-made forests of wood: buildings. Where there's construction, there's wood. No, I'm not suggesting that you tear them down. Well, I'm not openly advocating it, anyway. Rather, speedy construction means waste. On most sites, the builder/yardman must pay someone to haul away this construction debris and will, in most cases, be more than happy to let someone have it. There's a catch: you have to haul away all the other stuff, too. Since there are lots of other neat things to be had from this job, it's worth it. Do a good job and you may have a supply set up for some time. There's one drawback to doing this. You are, in the highest probability, dealing with developers, exploiters, and land ravagers. If your principles are high, pass it up.

Wood can be had from old houses that are being torn down. Or, if one is going to come down, you may be able to underbid everyone else and do it yourself. Lots of other goodies are to be had—electrical wire, fixtures, plumbing, sinks, glass, etc.—but be ready to invest a lot of time and hard work before you commit yourself to the job.

Another wood source is a tree that needs removing because it's going to fall down on a house or power line, or looks bad, or is diseased. Just start walking around and spot them and contact an owner. If a power line is threatened, call the utilities and offer to remove the offending tree. In neither one of these cases will you be able to just take your chain saw to it and let it fall; you'll do what you said the tree would do unassisted. But, with skill and cunning, the tree can be guy-wired to come down like a tower. Faster, yes, but where you want it to go. A controlled crash, it's called. Don't undertake this unless you know what you're doing or are heavily bonded. If it can't come down intact, you climb and cut—a job not recommended for anyone afraid of shaky perches. Hindsight is worthless here; if you can't envision (perfectly and every time) where the branch will go, find another source of wood.

Your own wood. You may be fortunate enough to have bought a piece of land that has an abundant wood supply on it. Even if you don't use any of it for building, you don't have to cut down the finest just for wood heat. Trees diseased and damaged (by snow or lightning) may be removed first. As well, stands of trees usually need thinning, which is preventive cutting to minimize stunted or deformed growth and to give some of the trees a chance to grow up big and strong. Deformed trees go next. Even this may be unnecessary. Heavily wooded areas often have plenty of downed wood or broken trunks, limbs, and branches, requiring only cutting up instead of cutting down. Finally, there are trees that threaten houses, barns, or electrical wires; no matter how fine they may look, if you don't cut them down, you take your chances. Or, work at moving the wires, buildings, etc. to another site. I'm serious. A falling tree should not be taken "lightly"; that's certainly not the way it falls!

You'd be lucky to get a cord of downed wood per acre or two, on the average; with heavily wooded areas or many accessible acres, this will be sufficient for a time. Eventually, woodlot management will become a necessity, but the effort expended pays more than just wood. First, you plant trees where you want them. You might as well make it accessible for a vehicle; removing the wood can sometimes consume more energy than cutting it down, cutting it up, and splitting it. Or plan it for a windbreak, cutting down heat losses in the winter and controlling snow drifting. Or plant it as a screen, cutting off a view you don't like, the neighbor's settlement, or prying eyes.

A second advantage of the planned woodlot is that it can raise the humidity and, with a good wind, present you with a refreshing breeze. A third benefit is that trees and underbrush provide shelter for fowl and beast alike, avoiding the manicured lawns and fields that characterize so many "country" places. Ever wonder what bothers you so much about them? It's the sterility of missing wildlife. A final benefit is that trees look nice, adding beauty to your life. Some people look at trees and see empires. Some people look at trees and see . . . trees.

TERMS AND ENERGY YIELDS

I've used the word "cord" a few times in reference to wood. A cord is a measure of wood and gives a quantity/volume reference, like gallon. The amount of wood that would fit into a space that's 4 feet long by 4 feet wide and 8 feet long is a true cord. That figures to 128 cubic feet of wood. Other terms, like "face cord" or "short cord," are used for convenience by firewood salespersons, and you should know what they mean before you buy or you might not be getting such a good deal. This is not like selling milk by the half-gallon to make the price seem lower. The lengths of wood cut for a fireplace will be different from lengths intended for use in a stove, oven, or wood-burning water heater. You can always measure the length, width, and depth of a pile, multiply these together to get the number of cubic feet of wood, and compare it (and the price) with 128 cubic feet.

A cord doesn't define solid wood. Look at the stack, and you'll see some air spaces in there. Bigger chunks of wood will have less air space around them than a whole bunch of smaller ones, but don't use this as a criterion; if the bigger stuff is selling for less, it's because *you* get to do the splitting. In practice, two-thirds of the cord is actually wood and the remaining third is air.

How much energy do we get from burning wood? That varies with wood type, the amount of moisture, and the amount of resin in the wood. Just so you don't think I'm trying to avoid the question, I'll give you an answer. *If* the wood has been properly dried for six months (as it should be for any use) and *if* we were to allow only a minimum of resin, we could expect to get between, for the worst and the best, 12 million and 24 million Btu's per cord of wood. Not exactly chickenfeed, is it? That's equal to the energy yield of a ton of coal, or a little more than 200 gallons of heating oil.

When you're gathering your own wood, you

don't have to stack it into cord quantities. There's nothing sacrosanct about that measurement unless you're trying to compare it with usage elsewhere, sell it, or buy it. It is, however, useful to know how much wood you're consuming each winter (or through the year) so you'll know how much to cut next year. You will have to stack the wood if you want it to dry correctly, so do it squarely or in a nice rectangle so you can compute the volume. Once you've got the pile volumes totaled, divide by 128 ft^3 and your answer is the number of cords you've cut and stacked.

TYPES OF WOOD

The many types of wood are split into the main classes—hardwoods and softwoods. Ash, oak, birch, maple, and cottonwood are some of the hardwoods, and cedar, fir, pine, and redwood are some of the softwoods. Hardwoods are deciduous trees; softwoods are evergreens. What you'll use is what's available to you, but if there is more than one type of wood, you should learn which will give you the most return for your effort. A forestry office should have pamphlets on the types of wood and the desirable and undesirable traits of each. Then, with a little course in tree identification, you're ready to stalk the woods.

The energy yield of the wood is of primary concern to the discriminating woodcutter. Since the worst wood gives only one-third to one-half the heat value of the best, time spent in selection is more than amply rewarded in less energy expended down the line. Generally, hardwoods give more heat than softwoods. Cottonwood, a hardwood, has less heat value, however, than Douglas fir, a softwood, so there are exceptions.

Other factors govern wood selection. How fast a wood burns may be important to the application you've in mind. Hardwoods burn slowly and evenly, whereas softwoods burn quickly and, mainly due to the resin they contain, sometimes very unevenly. Hot and fast burning is undesirable in any stove because of the damage it may inflict on the unit and the high heat loss up the flue. But starting a fire is also important. A hardwood rarely starts quickly or easily; it needs the rapid, hot burning of a softwood to ignite it. Therefore, don't be afraid to mix the woods you get and use. When you're freezing, having to light and relight a fire is no joy; a softwood can provide quick relief.

Some woods, like Douglas fir, give off a lot of smoke. Most of this stems from the resins in the wood. Wet wood and incomplete combustion can cause smoke, too. Fortunately, woodsmoke is thought of as rustic and smelling it is considered enjoyable; it's really not considered pollution at all. Woods do burn cleaner than oil or coal, and they produce no sulfur-based irritants or pollutants; extensive use will give us smog-type layers but the soup would still contain less harmful ingredients than fossil-fuel use generates. Nevertheless, if you've got a hidey-hole and don't want the advertisement smoke gives, choose a nonsmoking wood, dry it thoroughly, and use a stove that combusts it properly.

To dry wood requires that it be split into pieces small enough to aid moisture in escaping. Since some woods split more easily than others, this may be a criterion in the selection of the type of wood. If you put more than four cords through a stove in one winter or if your time is really valuable to you, a log splitter can pay for itself quickly.

Combusted correctly, wood leaves little residue except ash; this can be used as fertilizer or for many other homemade products. Since it can clog the stove once a quantity of it is amassed, it must be removed periodically. Exercise care, however, in where you dump it, as it may contain still-glowing embers. A metal bucket, therefore, works better than a paper sack, particularly if you set it on the back porch.

Other types of fuel generally require space or containers for storage, but wood can be stacked outdoors to dry and very little, if any, cover is needed to keep it dry. Disregard this if you live in a rain forest; you'll need a top cover or shed roof. While a nearby woodpile is convenient if you need to pop out there for a load it's not considered a good practice to stack it alongside the house; besides the obvious fire hazard, one must consider the insects and other critters attracted to these wood condominiums.

WOOD-BURNING DEVICES

I've already mentioned some of the uses to which wood can be put—space heating, water heating, and cooking. While we're eyeballing each application, we'll discuss the not-so-obvious attributes (or shortcomings) of the devices used to contain the fire and perform the work.

Fireplace. Fireplaces satisfy our most basic urge with regard to fire: to see it. Originally, however, the fireplace was an in-house version of the cooking spit. They're in houses today as a throwback and because some folks need the "campfire" their busy lives won't permit them.

For all their charm, most fireplaces leave much to be desired. Improper design or infrequent use leads to billowing clouds of smoke inside the room when they're lighted, and every bit of radiated heat is needed to compensate for the irritating drafts they create. Efficient they're not—most of the heat goes up the chimney, since there's no intake control of combustion air. Furthermore, as houses are made and insulated more efficiently, a fireplace may actually starve from lack of air, or, if the room is really airtight, it will go out. Masonry fireplaces tend to warm the room less than their metal counterparts, since the former absorb heat and the latter reflect it; if you desire a heat lag once the fire goes out, however, the brick or stone will slowly release its stored-up heat after the metal unit has cooled.

Improvements have helped the fireplace to survive. Some designs have air corridors built into the chimney or metalwork to extract useful portions of conducted heat, whether the air circulates through convection or blowers. Special glass, designed to pass thermal heat, can be used over the fireplace opening, making a fire screen unnecessary and allowing combustion air control through adjustable vents at the base. A few fireplaces have an inlet from the outside so that combustion air doesn't have to come from the inside of the house. This greatly reduces the loss of room heat and eliminates the hot-front-and-cold-back feeling as you face the fire. In extremely cold climates, this won't work; the icy outside air "cools" the fire too much.

Closed-box heater. Serious users of wood heat rely upon the closed firebox, ranging from the Yukon stoves (made from 55-gallon barrels—see Fig. 4-1) to the Ashley and Riteway (complete-combustion, airtight units). Because combustion air is perfectly controlled, even medium-priced units are leaps and bounds above fireplaces in efficiency. The more perfect the control of combustion air (no air leaks), the better the stove. The best stoves go a step further, rerouting unburned gases emitted from the wood back to the fire itself. Heat-activated controls are added to maintain the desired combustion rate, allowing more air if the fire is dying and rationing it if it starts roaring.

The main drawback to the closed stoves is that you can't see the fire. If that's too high a price to pay for efficient space heating, buy one of the new units that offers either an open or, shutting the big doors, a closed firebox. One other common complaint is that there is no residual heat once the fire is banked or dies out. True, perhaps, of the smaller, less expensive stoves, but not an issue with the automatic, airtight stoves which can go six to twelve hours on one load of load. Unlike the masonry fireplace, the good closed-firebox stoves can be used throughout the night and there's the ease of starting that warmup fire in the morning, too.

As to the controversy between the high-cost stoves and the simple, low-cost stoves, it all comes down to the application. Harsh winters mean high stove use, and nothing can hold up like the high-priced units, especially if you have more than one room to heat. For unattended operation, high heat yields for your hard-won wood supply, and durability, there's no equal. Riteway and Ashley, up until

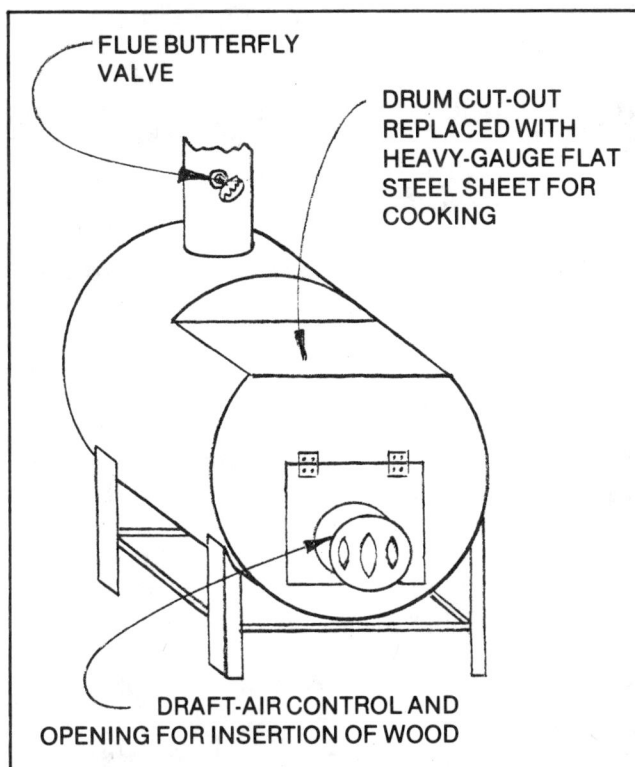

Fig. 4-1. A Yukon stove

a few years ago, were the best to be had, but with so many new stoves on the market they've got stiff competition and you've got some checking to do.

The simpler stoves do the job, too, if the job is occasional use, heating small rooms in the same house that wouldn't get heated from one big stove, or a desire to use something that you built with your own hands or to recycle certain materials. For the barrel variety of stoves, the Yukon design is the old standby, but there are others, still using drums, that match or beat its performance (see Sources and References section). If softwoods are burned in them, the thin sides will glow with the heat and will, in a short time, fatigue or rust away. Sand in the bottom of the stove helps to keep direct heat off the metal (let me tell you what a panic it is to have the bottom of your stove fall out while in use!), evens out the heat in the stove, and keeps radiated heat from melting your tiles or singeing the floor. Bricks may also be used for the bottom and sides, tempering the heat and keeping the glowing logs from resting against the metal. Even these simple methods will prolong the stove's life. It won't last as long as you do, but long enough to pay you back for your effort.

Cookstove. Cooking and wood heat are a good match. That you already know if you've stuck a pot of water to boil or a stew to heat on your space-heating stove. Wood-burning cookstoves do the job a lot better. First, they provide a larger flat surface on which to put pans, pots, and skillets. Second, while setting a pan on hot metal is all right, direct exposure to the flames maximizes heat transfer and wood-fuel efficiency. With this in mind, many cookstoves have removable circular sections so the pot is supported but still gets flaming heat. Some stove designs have concentric removable sections at each burner site so that you may put any size pot on it and still have it supported and exposed. Furthermore, the firebox is located on one side of the unit, with an oven alongside and burner plates directly above the oven and firebox alike. In this arrangement, you can pick the degree of heat required—burners above the oven are cooler than those just above the firebox.

Slaving over a hot stove is no fun in hot weather, which is one of the reasons that dual-fuel stoves exist. That is, you can use wood and gas (propane or natural gas) or wood and electricity. People who've gotten past the first fumbling attempts to use wood in cooking food agree with old-time users: it's just as effective. But it is different. Just as much as heating with gas will be for someone that's learned on an electric, and vice versa. Additionally, many of these ovens came equipped with a "bread warmer," a cabinetlike box braced above the back of the oven and through which the flue is channeled. The heat radiating from the flue heats anything placed in the cabinet.

A common ailment of the wood-burning oven was the eventual burnout of the firegrate or firebox innards from the heat. So, while used ovens are in abundance, some repair parts are needed, and often no source is in sight. A good solution is to rebuild the firebox, maybe enlarging it to extend down into the ashpan and lining it with brick; this protects whatever metal is left and helps to even out oven heating so you don't get food burned on one side or raw on the other. This is the time to add that outside-air flue for combustion air. Since the cheaper stoves had a non-airtight chamber, this would help alleviate any control problem if a damper is added. One rebuilding scheme I've seen (see Fig. 4-2 and Sources and References section) even put an ashpan under the house, and the inlet-air flue doubled as a means of removing ashes from the stove.

Water heater. With a space-heating stove or a wood-burning cookstove, hot water is not difficult to come by. Stick a pan or pot of water on top and wait. Great for cooking or dishwater or giving a baby a bath, but irritatingly insufficient for your bathing or showers. Others have apparently had the same feeling and solved the dilemma in their own way.

For masonry fireplaces, specially built water-filled tube grates supply the means to get hot water from a roaring fire. If you thought about this before the fireplace was built, water pipe can easily be laid

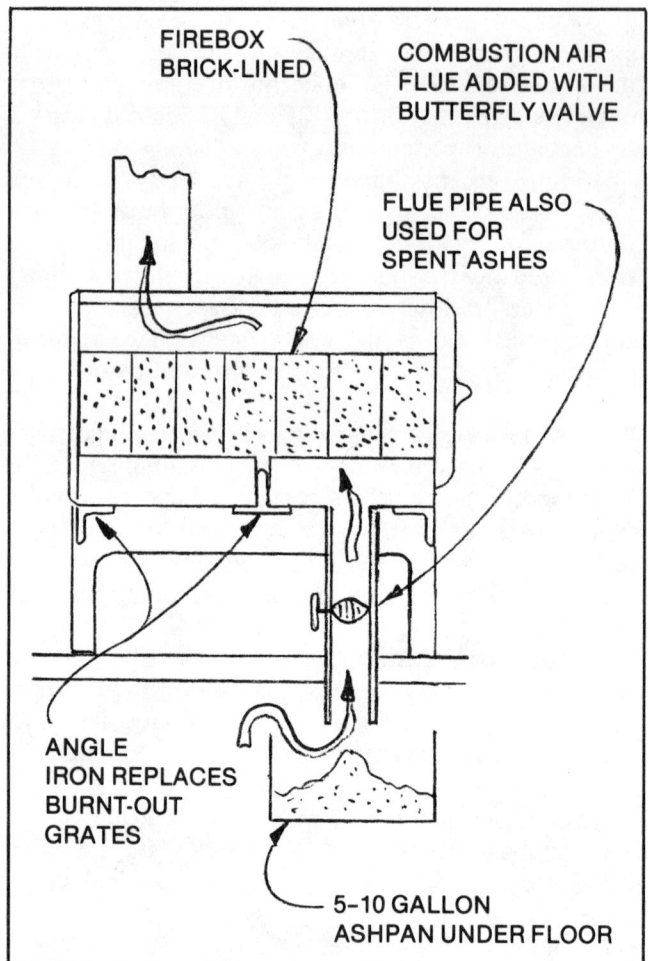

Fig. 4-2. A rebuilt wood cookstove

in the walls to absorb heat from the brick or stone. If it's designed well, a thermosiphon flow can store the heat in a tank for early-morning use or high-quantity needs. Otherwise, you run water through it when you need the heat. Secure the tubes well or you'll get a singing fireplace from the undulations the tubing will experience when the cool water hits it.

A few closed-box stoves have coils of tubing built-in for getting hot water; just hook up the tank connections and start a fire. Some neat designs have come out for coiling tubes inside the flue for heat extraction (see Sources and References section). A problem with any heat extraction in (or from) the flue is draft reduction; enough heat must always be left in the exhaust fumes to let them rise and exit the flue and building. As well, many of the provisions of a thermosiphoning solar water-heating setup must be enacted to prevent melting the tubes or a pressure/temperature buildup if the water flow is blocked.

If your hot water needs are high, build or buy a wood-burning water heater. Homebuilt versions come from discarded gas water-heater tanks or steel tanks; mine originated from an old depth-charge canister with a built-in flue up through the middle of the tank. A firebox is added, and with ash-removal, wood-loading, and combustion-air-limiting features incorporated, only the plumbing remains. Wood-burning water heaters are also manufactured; there are several American brands to choose from and many foreign ones. Of excellent quality are some of the Mexican water heaters (see Sources and References section). Most heat from 8 to 14 gallons of water. It doesn't sound like much, but with a handful (not an armful) of wood, you get 14 gallons of boiling-hot water in 15 minutes, all for around $100. That's convenience in style!

WOOD-BURNING PRECAUTIONS

Wood heat has its own idiosyncrasies, and you should be aware of a few of the more important ones. Since air is consumed in burning wood, we must always provide for combustion air; you've failed when you find yourself warm but gasping. Large quantities of CO_2 are produced in the burning of wood, and since we can't use this stuff along with most of the rest of the exhaust gases, we need to put it outside the house. So a flue is attached to the stove or oven or water heater, and this is routed through a wall or ceiling and roof. Or, rather, through *devices* which mount in the wall, ceiling, or roof. Don't do it without these, or the flue, which gets very hot, will burn down your house.

The stovepipe (flue) needs a damper to help control the combustion rate. When you buy the stovepipe, you get a butterfly damper and install it at a convenient height off the floor for yourself. In operation, it's opened enough to keep smoke from filling the room but closed to allow as few Btu's to leak up the flue as possible.

With wet wood, resinous woods, or improper stove settings, smoke occurs. Nice to look at, doesn't smell too bad, but it really gums up the flue's inside. The deposit is called creosote, and you don't want it in your stovepipe. Other than the fact that it's restricting airflow, interfering with additional space-heating through stovepipe radiation, and indicative of inefficient fuel usage (unburned combustants), it's also a fire hazard. If you get a hot fire going, it's possible to ignite this goopy mixture and have a flue fire. Ultimately, this burns down the house, because it melts the stovepipe and exposes the flame to all your combustibles. It's conceivable that you'd need the flues cleaned several times a season, but once a year is the recommended minimum. If you don't like to do it, hire a chimney sweep. This person rids fireplaces of the same hazard and has the right tools for stovepipes, as well.

If your stove sits in a corner or against a wall, a minimum of two feet should be left between it and the wall. Building codes specify materials that may be used in this vicinity; abide by that info. Avoid using asbestos if you're concerned about carcinogens. If you normally remove the stove before summer, reflective metal sheet will deal with radiated energy from the stove's back or sides, and is removable with the stove. Exercise caution in placing chairs, tables, clothes, etc., too near an operating stove; this also goes for the kindling and stacked firewood in the house.

Wood-heat users sometimes complain of the dry throat that results from burning wood. This is easily noted during long cold snaps when the house air is not fresh. A quick way to deal with this is to put a pan of water on any stove in use and let it boil away, putting all its vapor into the air. Boiling water consumes energy, so a preventive method is better. When you can't open windows to replenish house air, set out several containers of water to evaporate in different rooms.

WOOD-RELATED HARDWARE AND SOFTWARE

Unless you've got your own woodlot (planned or otherwise), you'll be heading off somewhere to get your wood. Getting your firewood permit is just the start; it's good for a year or less and needs renewing. At this time, no fee is usually involved. There are still a few good people who use hand-powered saws to cut their wood, but most folks use chain saws. And while some of these operate on 12 volts for use from your car's battery, most are gasoline-fueled engines. So don't forget the gasoline—the gasoline-and-oil mixture, that is, since these are two-cycle engines. This means that the engine gets its lubrication from the oil mixed *with* the gas; using regular gas alone will ruin it. Mix it beforehand; it's special stuff, so any old oil won't do. As well, different brands and sizes of chain saws use different proportions of oil to gas, and you must know this ratio before you mix your own or attempt to use some of Elmer's; his mix may not be your own. Mark the can so that neither you nor anyone else will confuse it with regular gas.

Chain saws use oil to lubricate the chain during use. That's what the other fill hole (or cap) is for. Check it frequently until you know how rapidly it's depleted and needs filling. Manufacturers have left enough of a reservoir so that, both starting full, you run out of gas first. But it's your responsibility. Running the chain saw out of gas is all right, or they'd have put a fuel gauge on it. Letting it run out of chain oil will burn up the chain, and there's no idiot light to warn you. After all, they'll be happy to sell you another bar and chain! Don't get all the way to your cutting site before discovering that you forgot to bring chain oil.

If you're cutting in a national forest, don't forget to bring your permit. If you're stopped, it gets downright awkward. At the least it's going to waste some of your precious time, and at most you get hauled in. Your fault. It says right there on the paper that it must be in your possession. Another thing: don't forget to get your chainsaw inspected and certified (at the forestry station or office) for an approved spark-arresting muffler. If you don't, Smokey the Bear will be angry. Even if you're not on public lands, this is a must.

Forestry permits require that you carry a shovel and ax, in case you start a fire, and it's a good procedure wherever you are. Acting quickly, you can prevent a major fire, but it's difficult enough without having even the simplest tools to stop it. Furthermore, a gallon or two of water will drown it past any resurrection later.

Chain saws need sharpening, and they get dull at the worst times. Even old-timers can't cut into a limb that grew around a rock, bolt, or wire, and that's that. A spare chain is a good idea; you put it on and keep going. As well, if you're doing a lot of cutting, you can have one in the shop getting sharpened and the other in use, rotating them. Or, better yet, get two chains and a sharpening file and learn how to do it yourself. Even if you can't eyeball the correct angle, you can buy a chain-sharpening jig that does it for you. At the price it'd cost you to get a chain sharpened, you'll pay for the jig and file in one or two seasons (depending, of course, on how much cutting you will be doing). Bring the file with you when you cut, for a touch-up sharpening when you're taking a break.

If you've never handled a chain saw, get someone who has to show you the ropes and then operate it in front of them. Invite them to cut wood with you and you'll gain more experience. There are a lot of ways to get in trouble; all but the freakiest can be spotted and avoided. Using a chain saw is fatiguing, and the noise and vibration will limit your cutting time. Take frequent breaks.

Wear the proper clothing, boots, and a hat. Wear protective eyeglasses, wide-vision ones, and replace them when they get scratched; otherwise, you just won't wear them. Gloves are a necessity, too. And, while it's last here, it's first on my list (having cut wood for years without them): earplugs. I can't overemphasize this. Buy the ones shooters use; they have an audio check valve in them that closes with sharp reports (like gunshots or chain-saw muffler noise) and opens for normal conversations. But even cotton is some relief. Otherwise you'll go for days unable to hear quiet sounds.

Once the wood's home, you split it. The minimum hardware is an ax, a sledge (6–12-pound, whatever you can lift and swing), and at least one wedge. Combination ax-and-sledge tools scare me more than double-edge axes, which I use as frequently as the single-edge. Having two wedges helps the game—and a game it is, too. You've got to learn what will split and how, and what won't. And you've got to get over hitting it timidly, or you'll be at it through all next winter. Control is just as important as strength, too. Or . . . use your log splitter.

FINAL COMMENTS ON WOOD

People who don't like setting up a wood heating system fall into four groups. First, there are the folks who, as children, had to gather wood, cut or split it, and haul it. They can't get over how much they hated doing this (forgetting, perhaps, that children hate to do most anything they're told). Second, we've got the people who've never used it. Whether from misconceptions or ignorance, they think it's got to be more trouble than it's worth; I hope I've reached them. Third, there are those who use it poorly, because they won't take the time or energy to learn how and won't experiment. What can I say? And finally, there's the group who don't like it because they don't have to like it and choose not to. That's okay by me. The way I figure, it just leaves more wood for the rest of us!

WATER POWER

It's unfortunate that waterpower suffers the same difficulty as geothermal, tidal, wave, and hydrothermal energy: you've either got it or you don't. If you don't, shake your head and skip over to the next section; due to certain biological functions, methane possibilities are universal. However, if you've a mountain stream that rambles through your land or a strong spring, there's a good possibility that you may be able to extract some useful energy from it. This chapter will tell you how to find out if it's enough, the different uses to which it can be put, and what's involved—hardware, layout, and design—in using it. If you've got a good-sized stream or river flowing through your land, it's not so much a question of whether or not to use it, but rather—what's holding you up?

ENOUGH WATER

Before we look at any hardware or designs, first we have to figure out if you have enough water with which to do anything useful. Or, I'll tell you how to find out, *you* do it, and *you* figure out if it will take care of your needs. Three things determine how much work can be harnessed from a flow of water:

the rate of flow, the amount of fall, and the seasonal variations. Let's tackle them one by one.

Flow rate. Flow rate is expressed in cubic feet per minute or second or gallons per minute or second. Knowing the flow rate is essential to sizing the system or figuring out if you should bother with a system in the first place. Flow rate is a product of the water's velocity (ft/sec, gal/sec, etc.) and the cross-sectional area of the streambed. Fortunately, using some basic hydraulics, we don't always need to find the cross-sectional area and water velocity to get the flow rate. I say "fortunately," because it gives you a few options in measurement techniques. There are three basic ways to measure flow—capture, weir, and float/section—and they're described in detail in Data Cubbyhole, Section G.

Seasonal flow. Rivers, streams, creeks, brooks, and other flow ways for water all vary in flow rate through the year. Flow is usually at its lowest in the late fall, and highest after the spring snow melts. *When* things happen isn't as important as *what* happens: there's more water. What you do with the additional flood is your decision; you don't want to leave it up to the water to decide for you. It will be more than happy to carry off your entire installation downstream. Measuring the flow (rate) of your waterpower source should include the maximum *and* minimum rates. Independent of the method you use to calculate the flow rate, you'll need to take readings several times in the year. An annual chart with readings every month or several times per month will help you plot energy availability during different portions of the year. If this is going to become your bread-and-butter energy source you may not want to wait a year. In that case, you must make adequate provisions for excess water, and you run the risk of a gross mismatch between the water you have, the power you can get, and the power you do get.

Head. Head is the measure of water falling vertically from one point to another. It's measured in feet, and we always reference head with respect to the lowest point it can fall. Water takes the easy way and goes along the path of least resistance, and, in order to flow, it must fall. One definition of a stream's fall is the difference in altitude (measured in feet) between the point it enters your property and the point it exits. The reason that it's very important for you to know the water's head is that this ultimately (and very strongly) influences the type of wheel or turbine you'll use. Once you've peeked at that section, you'll quickly come to appreciate having all the head you can muster.

Farther down the line, if the dawn breaks and you suddenly realize you've got a gold mine of a stream, creek, or river flowing through your land, and you need to lay out a sluice or drop pipe (explained later), it might prove beneficial to have a surveyor come in to find the head precisely—that is, to within a few inches. But for now a figure that's close, give or take a couple of feet, is adequate. There are a few methods to use, the easiest of which is borrowing a local surveyor's transit, but an adequate way is the pole-and-level technique. It can be done with two people but with three it's just as accurate with less effort, and more fun. This procedure is illustrated in Section H of the Data Cubbyhole.

Head and pressure. Head is pressure. I've already said that head is fall, but, before it's fallen, we can consider the water to have a "potential" for fall, and one way that we can describe the potential is to use the word "pressure." Fill a container one foot high with water and you'd have a foot's worth of water pressure. Screw a pressure gauge calibrated in pounds per square inch (psi) into the container bottom, and it'd read .433 psi. You'll have to take my word for it, because there are few gauges that will read that accurately. If the column of water is two feet high, the pressure will be doubled, or .866 psi. In fact, it's a linear relationship, going up .433 psi for each foot of water in the container, column, tank, etc. For example, a 30-foot-high tank has a pressure at its bottom of (30 multiplied by .433) 13 psi.

How does this change when the tank is made larger in diameter? It doesn't. It stays the same? Yes. Don't confuse the weight of water with its pressure. Pressure depends on the water's depth and that's all. Fill a mile-wide reservoir 10 feet deep, and a tank that's 6 feet in diameter with water to the 10-foot level, and, at their bottoms, there's the same amount of pressure. Strange, huh? The weight of the water in the reservoir will be quite large in contrast to that in the tank, but that's quantity, not pressure.

What if we run a pipe from this tank to a faucet down the hill? The tank is 10 feet tall; a quick multiplication gives us 4.33 psi at the base. What's the pressure of the water at the faucet? We don't know. Isn't it 4.33 psi? No. Why? Because we've got a column of water in the pipe, too. Oh, well, in that case, let's see. The pipe is 100 feet long. Multiply that by .433 and we've got 43.3 psi. Add it to the tank's pressure, which is 4.33 psi, and we get 47.63 psi, right? Wrong. Why not? Because the pressure corresponds to head, and head is measured *vertically!* If the pipe drops straight down out of the tank, then our figures *are* correct. But it's at an angle. Say that it's a 45° angle. The vertical component of 100 feet of pipe at a 45° angle happens to be 83.3 feet (I got that using basic trigonometry).

Multiply that by .433 and you'd get 36.1 psi. Add it to the tank's pressure (4.33 psi) and you'd get 40.4 psi total at the faucet. Seems like a lot of numbers, but work out a few examples and you'll have it down pat.

Ten feet of water in a big reservoir won't give you more pressure than 10 feet of water in a tank, but it will maintain the pressure longer and supply much more water (energy) in the process. A 6-foot-wide, 10-foot-high tank contains roughly 2,260 gallons of water (I just calculated it), but when half that's gone, the pressure has dropped by half. On the other hand, you'd be hard pressed to measure the pressure loss of a measly 2,260 gallons from a mile-wide reservoir when there's 18,000 gallons in only 0.01 inch of its depth. So if it's important to have storage that doesn't vary much in pressure to the turbine or wheel, wide is better than high.

USES FOR WATERPOWER

A traditional use of water is the mill, for grinding corn and grains. Yeah, just like the Dutch wind-powered mill. But there are others. I can still recall my surprise when I saw, for the first time, an old photograph of a water-powered shop. You get so used to seeing electric motors powering lathes, milling machines, planers, saws, drills, grinders, and sanders that you forget those motors weren't around even a hundred years ago. But they had shops, and they needed something to turn all that stuff. And one of the sources was water. That's only one of several things water is good for. I can think of at least four: mechanical energy, the ram, plain old water use, and electricity.

Mechanical energy. By far the greatest use of the water's energy is mechanical energy. It's not difficult to understand why. It's the first step in removing *any* power from the water. And because of the inherent limitations of many of the devices which extract energy from the water, it becomes too inefficient to try doing much else. But that's okay; there are many uses for the mechanical energy produced. As described in the previous example, water can "power" anything that needs a rotating energy source. With turbine or waterwheel, the center shaft rotates and so do the pulleys, gears, or sprockets coupled to it. Whether it's a whole shop full of equipment or your washing machine, it works fine. And, unlike wind used in these applications, a flow of water is incredibly constant and reliable for such usage.

Hydraulic ram. While different in appearance and operation from other water-energy-extracting wheels or turbines, the hydraulic ram is a water-powered device (Fig. 4-3). And it has one function: to pump water. Now, that seems silly. Spending dollars to *use* moving water to *move* water around. Why not just let it be—moving by itself to these other places? Fine! But what if one of these places is higher than the water level in the uppermost portion of the stream on your land? Sure, you can use the mechanical energy from the turbine or wheel to power a pump, but this ends up being less efficient. So, we use something which is designed only to lift water: the hydraulic ram.

The ram works by letting a flow of water work to pump a small portion of that water to a higher position. In the simplest sense, we get a flow going through a drive pipe into the ram and, suddenly, shut it off. Water, once moving, doesn't like getting

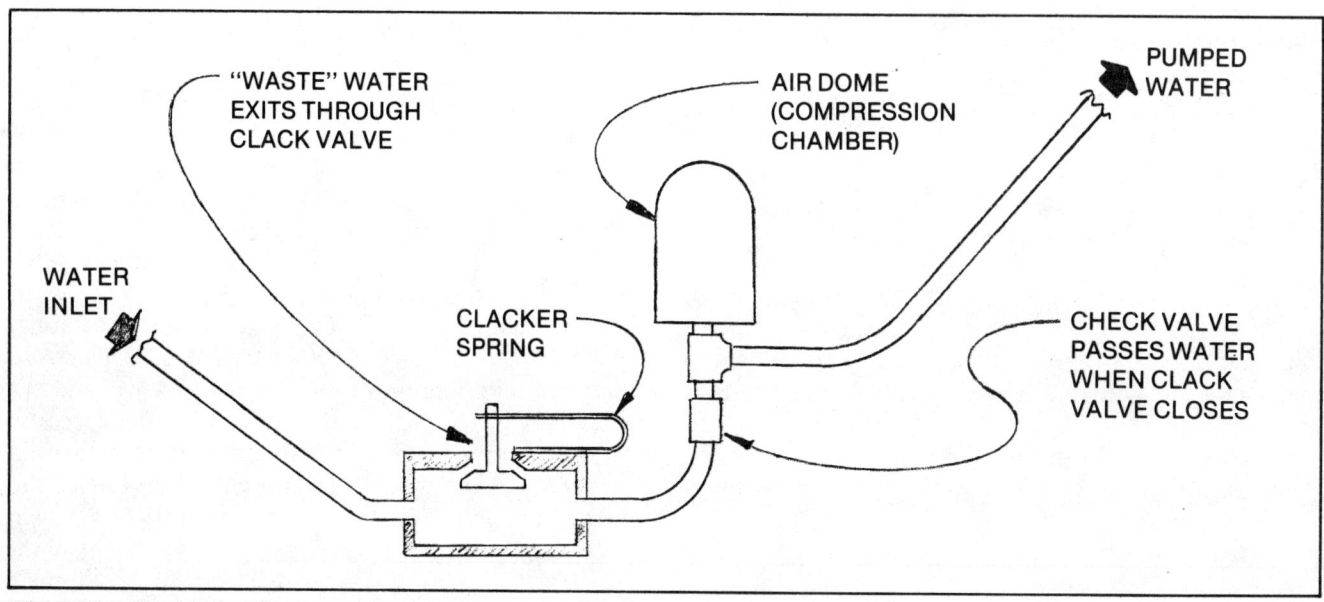

Fig. 4-3. The hydraulic ram

cut off so abruptly and so it piles up. And because it's virtually incompressible, it builds up pressure. If we put a check valve in the chamber, the pressure will pop it open, moving a small amount of water into the vertical pipe beyond. Once the penned water has spent its pressure, the check valve closes and (automatically) the flow resumes. Preset adjustments again shut off the flow and the pressured water acts upon the check valve. The water in the pipe behind the check valve climbs higher and higher with each cycle. You can attach extra sections of pipe until the suddenly blocked water does not create a pressure sufficient to overcome the weight/pressure of the water in the delivery pipe and deliver any more water through the check valve. That's the limit of the ram, and it can only be increased beyond that point with a larger inflow of water (diameter of drive pipe) or a higher pressure of incoming water (initial drive head).

Theoretically, the ram pumps a tenth of the water ten times as high, a fifth of it five times as high, etc. As we might suspect, in practice, the results are much less because of friction in the working parts—valves, inlet and delivery pipes, etc. Nevertheless, the results are impressive and beneficial if you want to fill a reservoir or get water to your homesite on the hill from the stream in the canyon below. If you've got gross amounts of water in the stream or river, you can use the hydraulic ram to pump water to an elevation and then let it drop immediately, or on command, into a water turbine that's back down the hill. Sort of roundabout, but undeniably practical under the right conditions.

The hydraulic ram is manufactured worldwide and the units are simple and easy to maintain and operate; they are, however, quite expensive. Owing to its simplicity, a multitude of homebrew ram designs exist for the owner-builder or person with a few shop connections; see Sources and References section.

Household use. Water gets taken for granted. Flowing out of the pipe, faucet, or showerhead, it's just too easy. It doesn't matter if you installed the supply system yourself; if it works okay, it usually goes unnoticed. But it does so much! Healthy bodies. Green growing things. Sprinklers covering an area that would take hours of time to water by hand. Washing away dirt and stains from clothes, dishes, counters, and floors. Maybe a pump gets it to us or maybe gravity does the job, but it has power of its own. Life itself, as we know it.

Electricity. A relative newcomer to the waterpower scene is generating electricity. On a large scale—Niagara Falls and Hoover Dam—it's been going on for a long time. But on an individual farm, or for a small community, it's practically brand-new. Why has it taken so long? First, the devices that extract the energy from water were designed for mechanical-energy uses. Second, most any use you could think of only a few years ago could be done using mechanical energy. Third, there wasn't any need to do it with water, thanks to fossil fuels. And fourth, if you had the need, you didn't have the information on how to go about doing it. I'll tell you what. You work real hard at the answers to two and three. Fair's fair . . . I'm whittling away at numbers one and four.

Electricity from water is not all that different from electricity from the wind. In fact, you'd find that the power formula is identical. Only two things change. One is the speed. Wind is measured in miles per hour, and water in cubic feet per second. If we combined the aeroturbine's area with the miles per hour of windspeed, we'd have cubic feet of air, too. Nonetheless, water is slow stuff, and wind is quick. But here's the second difference. Water may be slow, but it more than makes up for speed in density. Boy, is it dense! Water has about 800 times the density of air. Get hit with a bucket of air (without the bucket) traveling at 20 miles per hour and you'd scarcely notice. A bucket's worth of water at the same speed will, at the least, knock you over.

If you're lucky enough to have the right combination of head of water and rate of flow, even a small amount of electricity can be worthwhile. Doesn't sound like much to get only 300 watts of power from your stream, does it? But you cannot forget that it will deliver this, hour after hour, day after day, and if you're lucky, month after month. At 300 watts, for example, you've got a solid .3 kwh per hour, 7.2 kwh per day, 50.4 kwh per week, and that's over 200 kwh per month. Nothing to shrug off! If you could keep your monthly consumption and daily average compatible with this rate, you wouldn't need batteries! That even-flowing stream *is* your battery!

But we're getting ahead of ourselves. We can't go spending watts or horsepower we don't have—or maybe can't get. Let's peruse the machines which wrench the energy from water and put it in our hands.

WATER ENERGY MACHINES

Forgetting the hydraulic ram, everything that extracts energy from water for useful work can be lumped into two classes: the wheel and the turbine. And, while exceptions can and do exist, the wheels are slow and big, and the turbines fast and small. Across the board—delivering the same power—the physical-size differences of each type are quite pronounced. Within each group, we have further distinctions, and those we'll get to now.

Waterwheels. Slow and big, huh? Conjures up images of ponderous might. Reminds me of part of a poem by Francis Thompson *(The Hound of Heaven):* "... but with unhurrying chase, and unperturbed pace, deliberate speed, majestic instancy ..." No words can better express the feeling I had when my eyes first fell on a 30-foot-diameter waterwheel driving a millstone.

Waterwheels are further divided into three types: the undershot, breast, and overshot. Somebody thoughtfully named them according to where the water supply entered the wheel, and consequently our memory is not taxed. The "shot" is to the feet (under), higher up (breast), or to the hat (over). This also says a lot about *when* each was used. The undershot can handle a minimum of 1 foot of head but enjoys heads of 6–12 feet. The same upper range goes for the breast wheel. The overshot doesn't like anything less than 6 feet of head and prefers the 10–30-foot range. Only one undershot—the Poncelet wheel—and the overshot are worth further study.

Poncelet. The Poncelet wheel (Fig. 4-4) is the most efficient of the undershot wheels (60–80% efficient). It works from the impulse of water on its curved vanes. Its useful range of head seems to be 3 to 10 feet, but anything over 6 feet results in too large a wheel and an overshot may be more practical. A Poncelet wheel sits behind a small dam with a gate in it which can be lowered or raised to allow less or more water (respectively) into its vanes. Water flows into the vanes, riding up into the special curve and transferring energy in a slow and efficient manner. Shock from the water's impact is undesirable, as it robs efficiency. Due to design considerations, wheels under 14 feet cannot be made efficiently; irrespective of the head, this is the smallest size recommended. Furthermore, with increased head, the wheel size must be increased significantly until, with a 6-foot head and a flow rate over 20 ft^3/sec, the wheel may be as large as 24 feet in diameter. A good rule of thumb is: the diameter of the wheel must be at least twice, if not four times, the head. Framework for the Poncelet wheels is usually wood, and sheet metal forms the curved vanes. A 14-foot wheel of this type will sport over forty vanes.

An efficient Poncelet requires a breast. This is

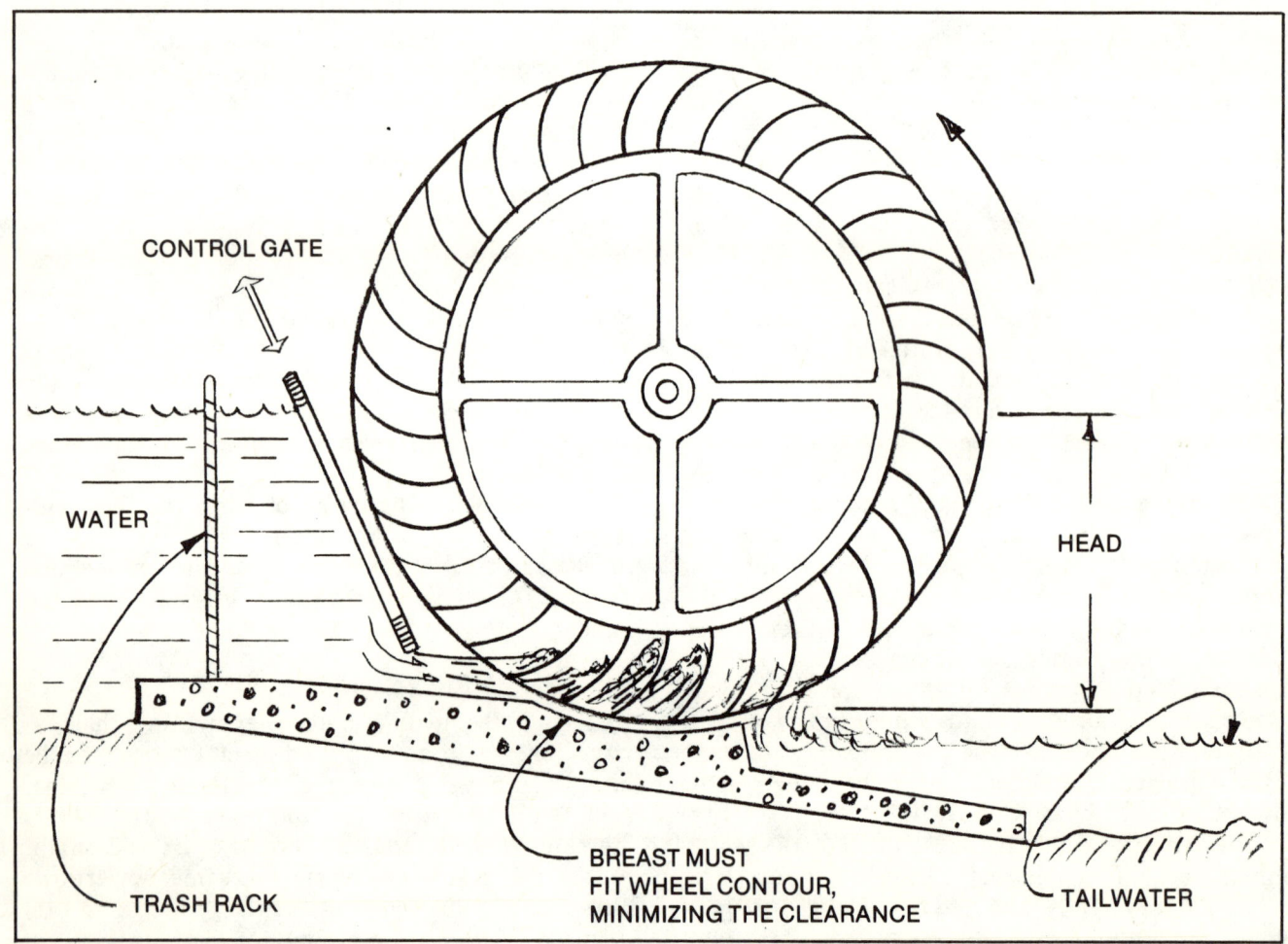

Fig. 4-4. A Poncelet wheel installation

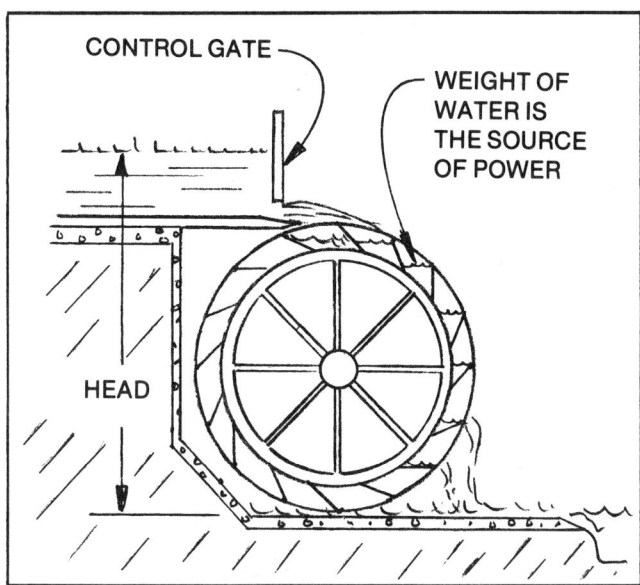

Fig. 4-5. An Overshot wheel

a close-fitting channel built as part of the damworks and positioned directly under the wheel. The idea is to guide the water to the vanes and not allow it any other means of escape from its duty. Herein lies one of the failings of the Poncelet: it's vulnerable to damage from small rocks, debris, or branches that get by the "trash gate" and jam between the breast and wheel.

In conclusion, for heads around 6 feet, the Poncelet is a good choice, giving very high torque. Its low rotational speed rarely exceeds 10–12 rpm and may be as low as 3–4 rpm. Mechanical energy seems to be its best application and only rarely will it be geared the required amount (200–300:1) for electricity generation.

Overshot. The overshot wheel (Fig. 4-5) takes over where the Poncelet wheel leaves off, working with heads as low as 6 feet but more comfortable in the range between 10 and 30 feet. It works best when it is "propelled" by the *weight* of the water, rather than its velocity (its impulse). Like the Poncelet wheel, the overshot wheel has curved vanes, but "buckets" would be a better term. Fill the buckets on one side and the wheel "falls" around in that direction. Except that the water is quickly dumped at the bottom, and the buckets come back up empty on the other side for a refill. Water comes in from the top through a sluice gate, which is adjustable; this is set to fill each bucket one-half to two-thirds full. Less is less power, and more gets everything wet, not to mention wasting water.

Overshot wheels are smaller than the head of water (measured to the tailwater); this makes sense because their water comes from directly above the wheel over the centerline. To make efficient (60–80%) use of the water, though, their diameter must be as close to the head as possible. Therefore, a 34-foot head will usually mean a 30-foot wheel. A 15-foot wheel in this circumstance would be grossly inefficient, using only half the head's potential. This factor imposes a restriction on the size of the overshot wheel; the size and resultant costs of construction and materials normally prohibit anything over 30 feet in height. Besides, there's close competition with the efficiency of turbines at heads above 15 feet without the enormous expense and huge size. If aesthetics is a criterion, however, I'd rather look at a majestic 20-foot wheel than the small turbine that would do an equal job.

Overshots get less damage from debris, and it's one reason for their popularity—they work well for a long time. Like the Poncelet wheel, however, they're slow-turning and require extensive gearing to use for generating electricity. Their awesome torque is suitable for mechanical work, and, made from wood and metal, they're simple enough in design and fabrication techniques to be constructed by the owner-builder; see Sources and References section for publications giving design and construction info.

Water turbines. Small and fast! The turbine stiffly competes with the waterwheels beginning with as low as 12 feet of head (for one type) and taking over the market beyond 30 feet if construction materials and cost are a criteria. A host of types come under the heading of water turbine, but we must dismiss from discussion those which are not suitable for an owner-built or owner-operated system. That leaves us with the Michell (or Banki) and the famous Pelton.

Michell turbine. Of the two turbine types, the Michell (Fig. 4-6) is the owner-builder version: nevertheless, efficiencies of 60–85% can be achieved, which doesn't exactly make it a clunker. Furthermore, it's the only machine that can deal with heads between 30 and 50 feet. It will work at heads as low as 6 feet but prefers 15 feet at the low end and rarely exceeds 150 feet of head. It's small (10–15 inches in diameter), works well at varying heads, turns at speeds between 200 and 2,000 rpm, and is able to drive a generator directly. Finally, we're talking about a waterpower device that can generate electricity without immense gearboxes and high gearing needs.

The Michell turbine blades are very similar in appearance to the Poncelet and overshot vanes, with one important difference. Water flowing into the turbine "impulses" the blades, moves through the turbine interior, and scores an additional impact on the opposite side before flowing out as waste water.

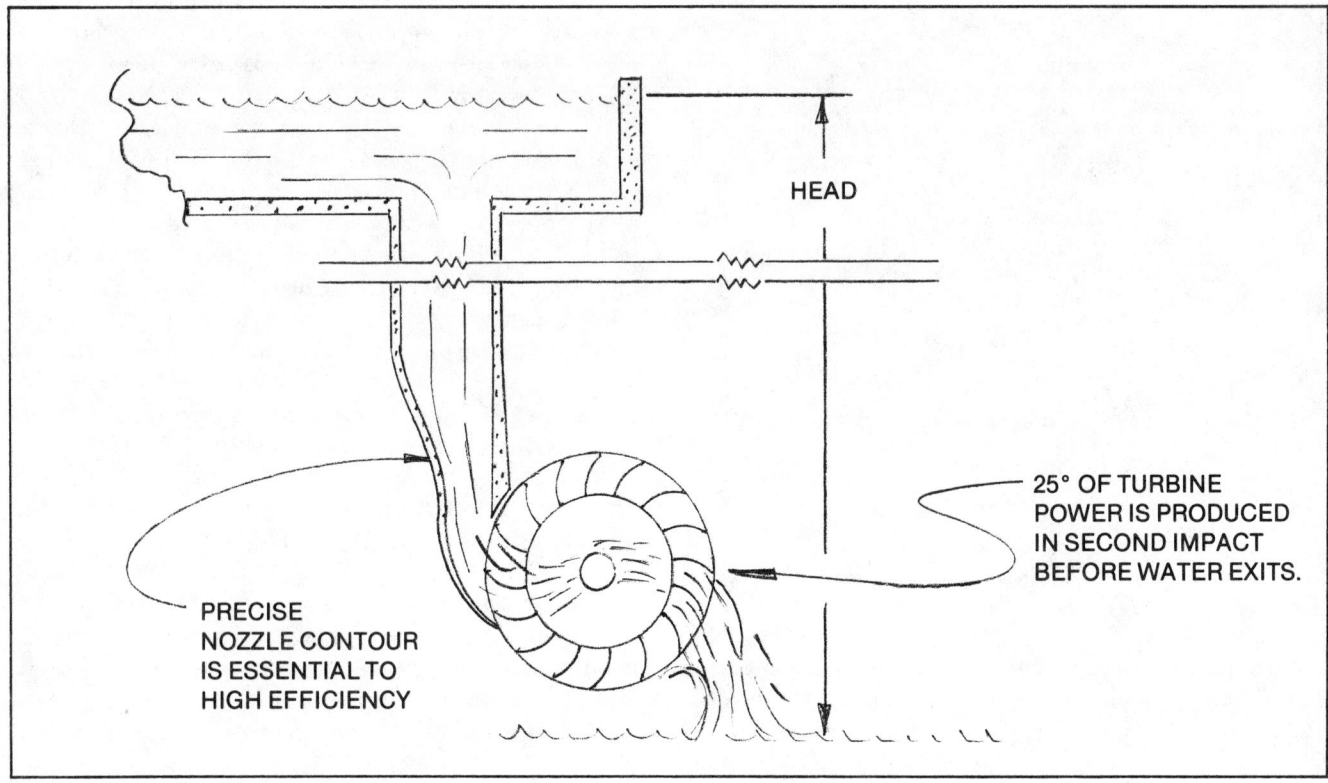

Fig. 4-6. A Michell (Banki) turbine

Approximately 75% of the energy is extracted as the water enters the turbine and the remaining 25% as it exits. Hence the name "crossflow turbine." Blade thinness is as critical to an unturbulated flow as blade angle. An additional boost in performance is accomplished by the unique shape of the nozzle which prepares the water source for entry into the turbine; it is one means of identifying the turbine very readily. Nozzle and turbine alike are all metal. Don't let it scare you, though; the unit is not complex to build, install, or use.

Pelton wheel. Small-scale generation of electricity from water and the Pelton wheel (Fig. 4-7) are a genuine marriage. An impulse turbine, the Pelton wheel is small in diameter (usually under 12 inches) and has small, bucket-shaped cups attached along the rim. Water is fed at high velocity through a nozzle at the cups, and it whirs. Rotational speeds as high as 1,000 rpm may be achieved. To increase the power output, the same wheel may be installed with as many as four nozzles, although two is the normal maximum you'd see anywhere. The Pelton wheel, like the Michell turbine, is enclosed in a steel shroud to prevent massive spraying of the immediate environment and to aid in removing the waste water.

Few Peltons are installed with heads of less than 50 feet, but the maximum head is well above 3,500 feet, making this an ideal choice in tapping mountain streams, with the only limitation being the amount of pipe you can afford. Pipe sizing is critical to minimize friction losses and, thus, effective head.

Due to the high rpm a Pelton wheel can attain, its construction and design are critical for efficiency (80-94%), with a major portion of the effort expended on the nozzle which feeds it. It's not impossible to make your own Pelton, but much more difficult than the Michell turbine for the same results. Nevertheless, if high power is needed and a very high head is available, its performance cannot be topped.

POWER FROM WATER

Gee, we haven't even gotten to horsepower and kilowatts of power and already we've been swamped with head, pressure, rates of flow, torque, rpm. Where does it end? Good question. The answer? Right here, in this section. It doesn't matter whether you're deciding to use the water for mechanical energy or electricity, this is where we crank out the answers.

Horsepower. If you're planning a small-scale hydroelectric installation, you don't absolutely need to deal with this subsection and the formula it introduces. Soon I'm going to provide you with a formula that will give a direct kilowatt reading from the data you plug into it. However, it's sometimes nice to know how much horsepower we're dealing with be-

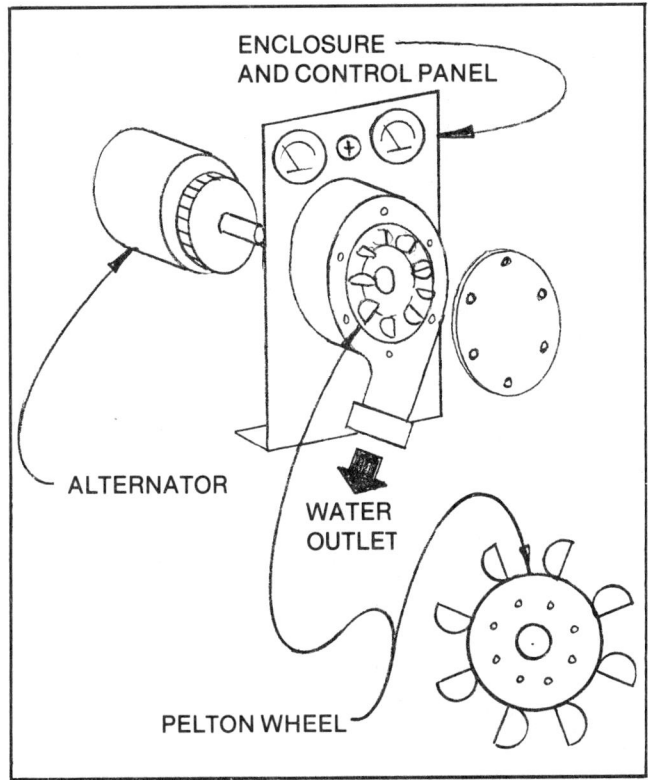

Fig. 4-7. A Pelton wheel-powered hydroelectric unit

Type of Wheel or Turbine	Diameter of Wheel Relative to Head (H)	% Efficiency
Undershot	3H	25-40
Breast	H - 3H	40-60
Poncelet	2H-4H (14 ft Min)	60-80
Overshot	3/4 H	60-75
Michell/Banki	1-3 ft	70-85
Pelton	1-15 ft	80-95

Fig. 4-8. Sizes and efficiencies of waterwheels and turbines

cause if there's any gearbox involved, it will need to be rated for the horsepower that's driving it. Put another way, you don't use an Erector-set gearbox with a water turbine; it'd just disintegrate. Also, once you have either horsepower or kilowatts, you can easily convert from one into the other.

If you want to get horsepower, or just think you do, you need to know the rate of flow in cubic feet per second (cfs). And you need to know head in feet. To use my formula, at least, by conversion, you can get a formula to deal with cubic feet per minute (cfm).

So, power in horsepower is:

(1) $$P = \frac{F \times H}{8.8}$$

where: F = flow, in cfs
H = head, in feet
8.8 = a conversion factor for the units to horsepower

A simple formula, and it's easy to use, right? Multiply the head by the flow, divide that product by 8.8, and you get horsepower. The only problem with that is that it's theoretical. It's the maximum potential of the water of that quantity and head. But our turbine or wheel is far from 100% efficient. Yes, most are above 50%, but thereafter, the efficiency differs between individual wheels and turbines and the quality of construction in each. Accounting for the difference is a simple matter. Let's just add an E into the formula and, for the turbine type and the precision of its build (Fig. 4-8), we'll enter that figure when we finally select one.

Head is another matter. When you go out and measure it, that will be a real thing, of course. But to use it to maximum benefit would be impossible; pipe friction alone will rob us of the *effect* of head. So, what we measure is the theoretical head (the maximum obtainable) and the effective head is what we can measure on a pressure gauge at the turbine. Now, where do we acknowledge that in the formula? Some people adjust H to compensate for the losses. Therefore, H is effective head. When you've got it measured, reduce it by 25% and your calculations will always be very close. Another place to put it is within E, along with turbine or wheel efficiency. If you combine the two, don't forget that efficiencies are *multiplied*. Therefore, a 76% efficiency in delivering head and a 75% efficiency of the wheel will give you a combined efficiency, E, of 57%. For our examples, I'll let H represent the effective head.

Now we have:

(2) $$P = \frac{F \times H \times E}{8.8}$$

Let's use our newfound friend. Our weir measurement of flow has given us 4.4 cfs (cubic feet per second) and we've measured the head at 24 feet. How much horsepower do we get? Before we start multiplying and dividing, let's adjust H. If we assume only a 25% loss of head (which is liberal rather than conservative), our effective head will be 75% of 24 feet, or 18 feet—that's H. What about efficiency? Well, due to our time-consuming efforts, we know that we've selected and built an overshot wheel that's at least 70% efficient.

Thus:

(3) $$P = \frac{F \times H \times E}{8.8}$$

$$= \frac{(4.4)(18)(.70)}{8.8}$$

$$= 6.3 \text{ hp (horsepower)}$$

That wasn't so bad, was it? This is the amount of power that's available at the center shaft of our wheel. As in the case of the windmachine, we know other taxes will be collected down the line—gearbox, bearings, etc.—but this is the shaft horsepower. If the flow figure you used is the maximum flow the stream develops, this answer will be the upper design limit of your system. It it's hard for you to visualize how much power that is, stay tuned for the next section. Few of us deal with horsepower enough to get a feel for it, but most everyone understands electrical power—watts or kilowatts—and we'll convert this figure next.

Electrical power. Generating electricity from the stream on your land involves the use of another formula. Well, if you look at it for a second you'll see it's not a completely different formula, but one that's got a teeny change in it.

(4) $$P = \frac{F \times H \times E}{11.8}$$

where: P = kw, (kilowatts), or 1,000 watts
H = effective head, in feet
E = efficiency of turbine or wheel
11.8 = conversion factor for the units to kilowatts

The only thing that's changed is the conversion factor—from 8.8 to 11.8—and E may change with the wheel or turbine selected. In this case, we've built ourselves a fine Michell turbine and know that we've got at least 70% efficiency (it's a proven design). But we've got a few more items to add to E. It'd be a shame to calculate the kilowatts at the shaft, as we did horsepower; that's just paper power. To my way of thinking, kilowatts is something you read on a meter. Or two meters—amperes and voltage—and multiply them together. Otherwise, it's just not real electricity. But the energy must pass through at least two more toll stations before it hits the meters—the generator or alternator, and electrical wires. Unless you've got facts and figures from the manufacturer, the generator could be as low as 50% efficient and as high as 95%; let's assume 65% and go on. Wires, if properly sized, will seldom fall below 95% efficiency. Combining them, we get:

E = (.70—Turbine) (.65—generator) (.95—wires)
E = .43

That's quite a chunk—only 43% efficient—and you'll want to do something about it when you select all those items to improve on it. But now let's look at our formula:

(5) $$P = \frac{F \times H \times E}{11.8}$$

$$P = \frac{(4.4)(18)(.43)}{11.8}$$

= 2.9 kilowatts, or 2,900 watts.

I'll take this stream any day; I wish it was on our land! This is real power that you'll get and, unless this stream varies in head or flow rate considerably, it'd satisfy a lot of applications. In a twenty-four-hour day, that's 69.6 kwh and in a month you'd collect over 2,000 kwh. A stream with one-tenth this flow or head could be considered seriously for the installation of a waterpower system; as is, it's a foregone conclusion.

Converting horsepower to kilowatts. It's sometimes convenient to switch back and forth between kilowatts and horsepower. There are 746 watts in 1 horsepower. Thus, .746 of a kilowatt (1,000 watts) is 1 horsepower. If we wish to speak in terms of kilowatts, we can say that 1 kw is equal to 1.34 hp. Maybe you can now see the difference between formula 2 and formula 4—8.8 and 11.8—because if we divide the first by the second, we get .746, and if we divide the second by the first, we get 1.34, further substantiating the conversion factor.

To convert, we have:

(6)   Unknown hp = (1.34) × (known kw)

or

(7)   Unknown kw = (.746) × (known hp)

To breed a little familiarity, let's convert the horsepower we calculated in formula 3 into kilowatts. Using the correct formula, formula 7, we have:

(8)   Unknown kw = (.746) × (6.3 hp)
                   = 4.7 kw

Again, this is at the shaft instead of what you'd read on some panel meters. Herein lies the danger in converting directly from hp to kw, or vice versa—you must remember at what point the power figure is being computed. That is, if we computed the hp of the 2.9 kw figure established in formula 5, we'd get:

(9)   Unknown hp = (1.34) × (2.9 kw)
                   = 3.88 hp

This figure is too low it it's supposed to represent the amount of hp available from the turbine, because contained within the original kw calculation are the losses experienced at the generator and electrical wires. This would goof you up if you were trying to "size" the gearbox for the maximum horsepower, because, using formula 4, we'd get the following for the factors used in formula 3:

(10) $$P = \frac{F \times H \times E}{11.8} = \frac{(4.4) \times (18) \times (.70)}{11.8} = 4.7 \text{ hp}$$

Between the answer in this calculation and the answer in formula 9, we've got quite a difference—almost 1 hp. Know what you're talking about and where it's coming from and you'll only have to worry about arithmetic errors, dropping a decimal point, or punching the wrong digit into your calculator.

Dams. Choosing the most efficient turbine is one way, although expensive, to get more power from the combined action of head and flow rate. On the other hand, with a varying head or flow rate, this course of action can backfire. A turbine or wheel designed for maximum efficiency often sacrifices versatility, its performance dropping rapidly with a reduced flow or head. To avoid "ping-ponging" on the turbine or wheel type, attack the problem from a different angle, stabilizing head, flow, or both. In other words, store some of the excess for use when flow or head leans out.

A dam fills these requirements, giving a reserve supply of water as needed. If it's built high, you can increase the actual and effective head of water available to the turbine or wheel. To stabilize the head, a low dam might prove effective for some types of machines that require it. If head is no problem but your site lacks a sufficient flow rate to be useful, a dam can gather water at a lesser rate and release it at an increased flow rate that's of appropriate size. A waterpower system that operates only fifteen minutes out of every two hours is better than nothing. Even if you don't have a means of storing the electricity (in batteries), you might be able to generate electricity as you needed it, say for a couple of hours in the evening, by using the dam as a "battery."

Dams are often used to help divert a flow into a turbine or wheel assembly. If a channel is used (sometimes called a headrace), the dam assures the waterpower installation of all the water it needs while spilling excess water into the stream. A control gate at the installation or a sluice gate at the cutoff point will allow shutdown or fine adjustments of the operation as needed. Whether for diverted flow, a stored reserve, head, or any combination of these reasons, a dam will do the job.

If a dam is dictated by your site, you'll have plenty of decisions to make—type of dam, where to put it, how to build it, whether to put it in the stream or out, how high, etc. This means engineering, and it's beyond the scope of this book to detail it. Check out the Sources and References section and, if you're talking about a big installation (above 5 hp), get some help. Your county Soil Conservation Service will often be happy to give you a hand.

In-stream dams suffer from the same slow calamity that is befalling big hydroelectric installations: silting. It's only a matter of time before reservoirs fill up and lose their storage potential. As long as fossil fuels are abundant, dredging the silt out will work, but it gets more and more expensive to do. Out-of-stream reservoirs, however, can be provided with silt or sediment traps (deep and slow-running "pits") to keep it out of the reservoir itself. Yes, you'll still have to dig it out, but it's not spread out as much (compared to the dam) and, if provided with a drainage method, the silt trap won't be inaccessible underwater.

Sediment and silt further muck up most installations by wearing down the exposed and moving parts. Ever heard of sandblasting? That's with air, of course. Well, tiny waterborne grit flowing in water is like liquid sandpaper. Open up a turbine that's been in use for some time, and you'll find it shiny and pitted. Since smooth surfaces are practically frictionless, efficiency is high. Pit it with the grit, and efficiency goes down. Pit it some more and the thin vanes (blades) will be damaged. Every little pit hurts! And they're going to take chunks out of your pocketbook to replace, too.

Fish in the stream constitute a problem if you build a dam; you may be required to build a fish bypass as part of the spillway. And speaking of spillways, don't forget one, and build it right. Without a spillway, your dam will overflow or bust. Overflow will weaken the foundation of a dam unless it's heavily reinforced to withstand the splash or turbulated tailwater. Reinforcing just one portion of the dam is cheaper and more effective; that's the spillway. After a lot of washed-out dams, they've got it down to a science and can do it right the first time. Why experiment? Find and use the info.

Channels or canals may be thought of as dug-in aqueducts, although, in waterpower terminology, you'll hear the word "sluice" (instead of "aqueduct") to define suspended channels. At the upper end of the channel is the dam or stream cutoff point. The channel's purpose is to get the water to the waterwheel site but to maintain the head in doing so. So, while the stream drops with the terrain, the channel hugs the hills, dropping just enough to keep the water moving along. At the end of the channel is the sluice which feeds the wheel; it's characterized by the sluice gate. Set into grooves in the sluice, the gate moves up or down to adjust flow. If a control gate is used at the wheel, the sluice gate serves only to roughly adjust water flow or to shut it down entirely. A spillway should be built into the end of the channel to avoid channel overflow during flood periods and to dump the unused water if the wheel is shut down.

Digging channels is a big job. Surveying is needed to assure a steady but minimal drop. Once the channel is dug out with muscle power, backhoe, or Cat, it must be sealed. A lot of clay in the soil might do this effectively, but you might need to add it or line the channel with a water barrier or concrete.

An alternative to the channel is pipe. If you're feeding an overshot wheel, it will be necessary to follow the land contours as with the channel. Turbines, however, are impulse-using, depending on pressure. That is, even if you pipe or channel the water to a point directly above the installation, the water can't come down a sluice. Rather, it must come down a pipe or some other *enclosed* space absent of air. You've got to keep them nasty bubbles out of there. I'm not kidding. Bubbles cause cavitation—a power-robbing, vane-damaging effect.

For turbines, then, a pipe carries the water from the uppermost part of the stream (still on your land) or any other handy route to the turbine's nest. Provided that the water that enters the pipe at the inlet has a flow rate equal to or greater than the amount that exits the pipe at the turbine, the pipe will remain full of water, and full head will be maintained—minus friction losses, of course. Larger pipe will reduce friction and increase effective head but, at some point, expense and gain must balance out. Get a library copy of a plumber's bible and work out pipe type, distance, and flow rates to the best combination for you.

Most folks that visit your place won't think of your pipe as a source of water so much as a source of eye pollution. For this reason, and others, think about burying it. If you don't, you must anchor it on steep grades or it'll come slithering down on you some day. It'll need support over dips and mounds to keep from busting from weight or shock; the joints are particularly vulnerable. Burying the pipe gives good support and prevents freezing or ultraviolet degradation (in the case of plastic pipe), and buried pipe is less susceptible to damage from people or animals. Get it below the frostline. If that's impossible, keep water flowing in it at all times; if you must shut down the turbine, allow the water to bypass at the base. Turning off the water at the base only keeps the pipe full of water. If it's likely to freeze, it's better to drain the pipe until the turbine is operational again. For some installations, the losses due to friction may pay off in the wintertime; coupling the friction's "heat" with the effect of fast-moving water, freezing of the pipe is unlikely. But if it's out in the open, the pipe may sag or crumple under the weight of accumulated snow and ice.

Any turbine or wheel is susceptible to damage from debris. If it doesn't physically damage the device, it may jam it. Or clog up the flow of water to it. Removing leaves, branches, logs, drowned cats, and other floating or submerged debris is the job for the trash gate. It's made as tough as the stuff that might find its way into your installation. Large-flow installations may need two trash gates—one for the really big stuff (logs, floating Volkswagons, etc.) and the other for finer items. Fish-supporting streams will need a mesh fine enough to stop the fish and wide enough (to reduce the pressure of the flow at any one point) to keep from pinning the fish to the screen until they die. The trash gate(s) will need cleaning; a larger-area screen will go longer between cleanings. It must be well supported in position lest both it and whatever ripped it loose end up going through your wheel or turbine. Channels and sluices will experience more of a problem than piped water, but it's more difficult to clear a pipe clogged with leaves than a sluice. The trick is, when using pipe, to keep leaves out of the pipe in the first place; a screened intake helps. A super-fine mesh or cloth will minimize silt or grit that gets sucked in.

Electricity storage. A dam will store electricity, in a manner of speaking. But our turbine will only put out so much power, and if we need more, extra water in the dam isn't going to give it to us. Here's where you make the decision. No matter how much power your unit delivers, it's either enough or it isn't. Or is it? It's time to rethink your rate of consuming power, making do with less. On smaller installations, however, this may be asking too much. You'll need to get some batteries (Fig. 4-9). It'd be redundant to tell you how batteries work in a water-power system, what kind to get, how to take care of them, etc.; it's all written down in the appropriate section in the chapter on wind energy. Take a look. It's all the same.

Well, not quite. A waterwheel is like a windmachine—if it's putting power into a load, it turns at a different rpm for a given windspeed than if it's not. When the batteries are in a low-charge state, everything's hunky-dory. When the batteries get topped, we need a regulator to reduce or stop the flow of electric current to them.

Now, who's kidding whom? Regulators are there to take care of full batteries. Who's got that kind of problem? I wish I did! I use every ounce of power my system is designed to deliver. I'm bordering on power poverty. Whoever heard of installing a system and using less than that amount? So, no voltage regulators and no current regulators—a straight hookup. I'm just waiting for that flash flood and then I'm gonna throw a party!

However, regulators are cheap insurance. It only has to happen once—needing to regulate voltage or current—and you don't know it or aren't around to do anything about it. Merely electrically disconnecting the generator from the batteries is a no-no. An "unloaded" turbine or wheel, still under the same rate of flow, will speed up. That's usually no problem for the turbine/wheel, but it may be for the generator or alternator. At 800 rpm, a 25% increase in speed brings the generator up to 1,000 rpm. A 25% increase

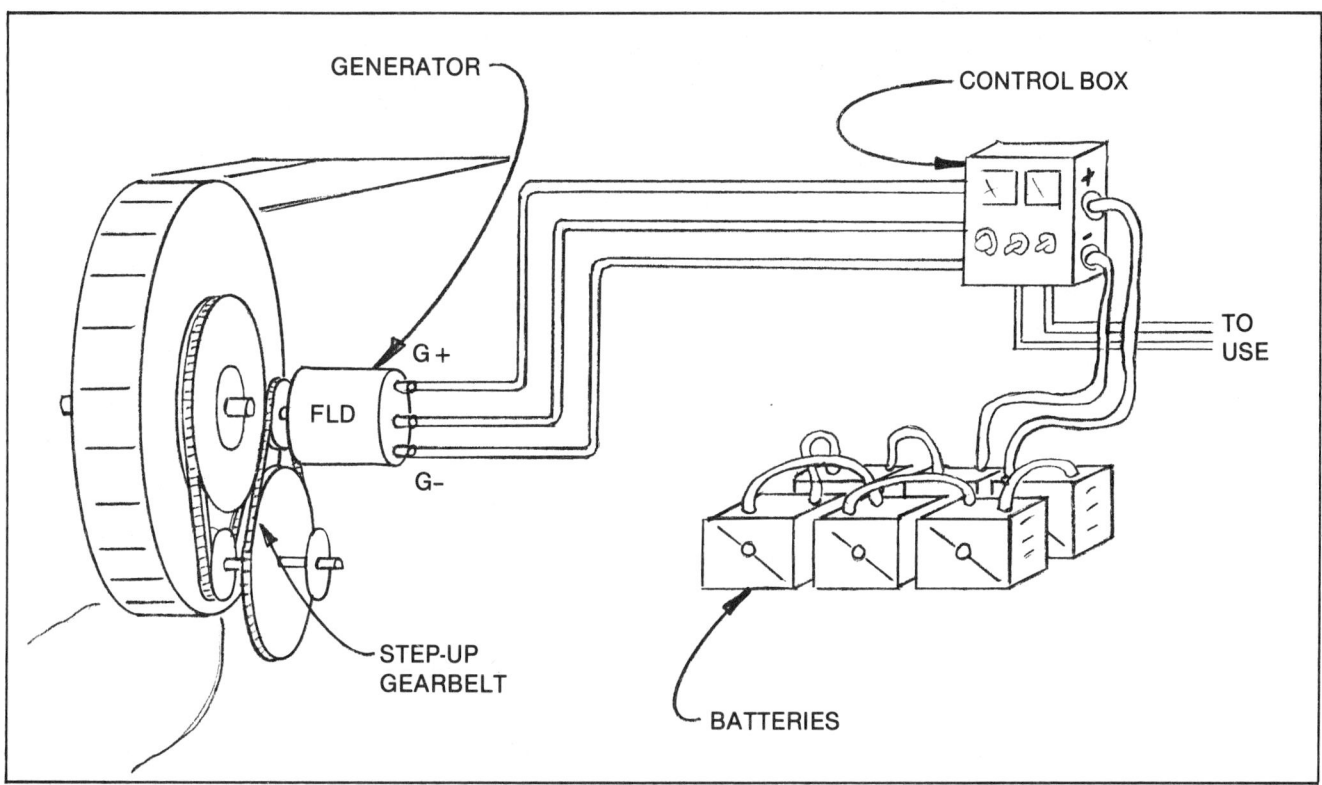

Fig. 4-9. Adding battery storage to the waterpower system

in an alternator's rpm when it's rated at 3,600 rpm snaps it up to 4,500 rpm. In my book, that's redline—say bye-bye to the alternator or its bearings.

A no-battery waterpower system has the same protection needs as the ones using battery storage. There's a threefold solution. First, disconnect the generator and alternator; automatic is better, but manual will do if you're a real homebody. This will allow the turbine or wheel and the gearbox (if used) to speed up, but they can handle it. Ignore this statement if you've got a high-speed turbine; it gets into the same trouble as the high-speed alternator. A double benefit from having the alternator/generator disconnect (clutch or neutral gear) is for emergencies.

Second, rate the equipment for the correct or a higher speed. Better yet, match the generator and turbine/wheel. Used and homebuilt rigs don't always permit this luxury, but I thought I'd mention it for laughs. If your turbine and gearbox are straining to begin with, it won't take much of a surge or emergency to push them over the edge.

Third, use a load diverter. When the batteries are topped, voltage-sensing relays kick in, and you start heating some water, giving a little more light to your greenhouse-caged plants, etc. Or just burn up the energy in a resistor. But keep the load uniform. No-battery systems need these devices most of all. Select all your primary circuits, with a relay on each. You kick on the light, you get the juice. Turn it off, it turns something else on.

I lied. There are six ways to keep your aeroturbine/gearbox/alternator from overspeeding. Number four is to shut the sluice gate (wheel) or gate valve (pipe). Five is to dump the water through a bypass; what a waste! And six is to throw a huge log into your waterwheel! Guaranteed: no overspeed.

Seasonal waterpower. Few are blessed with water flowing through their own land year-round. Many more have none at all. Then there's limbo—all the rest of us with our seasonal creeks, streams, or gully-washers. What a tease! To install or not to install? I'm of the opinion that some is better than none at all. If I can use wind energy (and I do), I'm ready for sometimes-there-sometimes-not energy.

Our own creek flows pretty well for about four or five months out of the year, drying to nothing for the rest of it. The way I see it, God is trying to tell me I need to clean the trash gate and remove all the accumulated silt and, knowing how busy I am, he gives me lots of time to do it. When the creek gets flowing, its rate of flow varies with the weekly rainstorms—not what you'd call uniform. First off, my wife and I figured that with all our wind-electric machines, we didn't *need* the water-made variety of electricity—*wanted* a nice waterwheel, yes, but didn't need one. But a reservoir would be nice to save water for the summer. The trouble was: where to put it? We had a nice site picked out, but it was above the

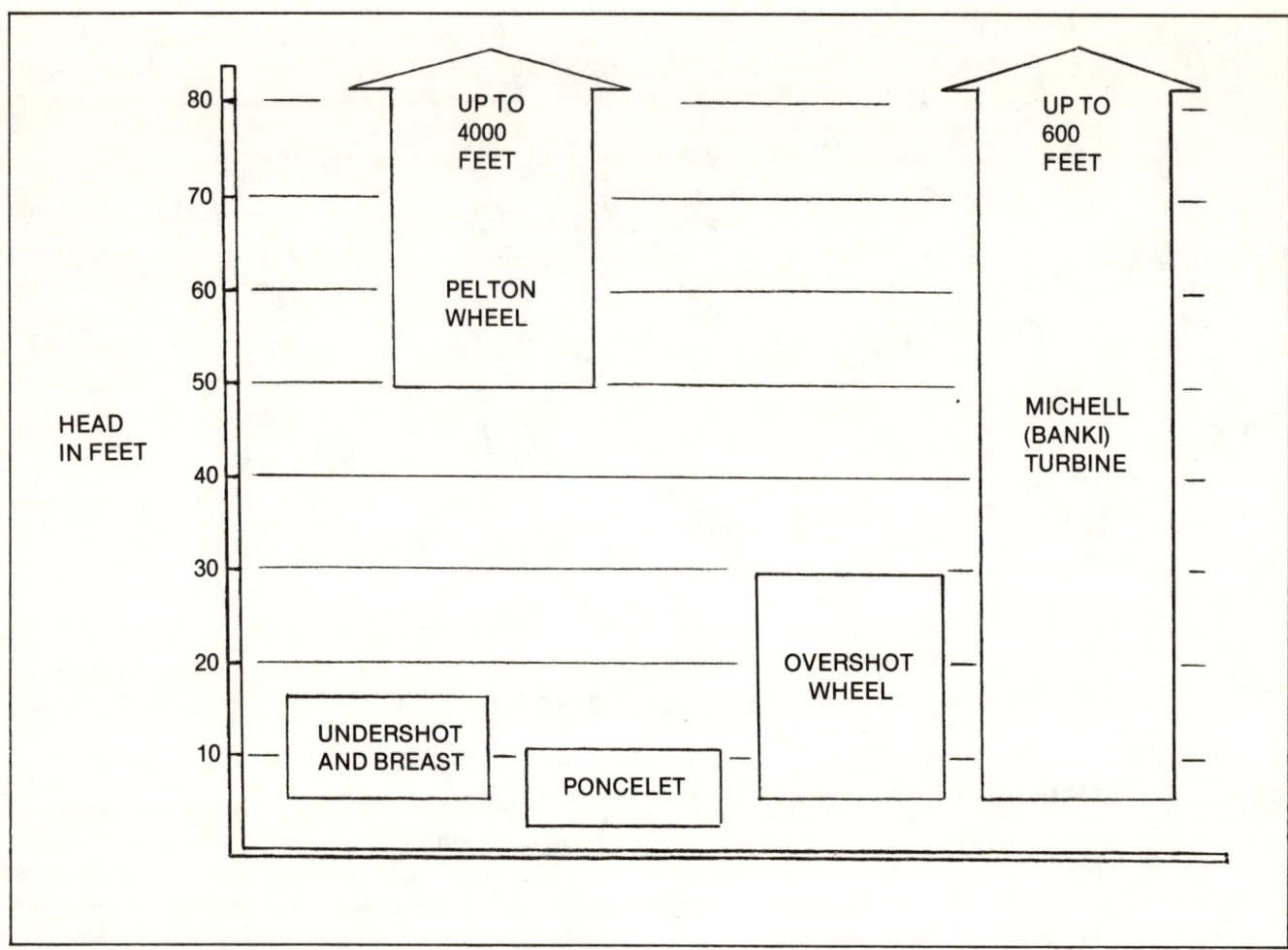

Fig. 4-10. Useful range of head for waterwheels and turbines

stream's entrance onto our land. How about a hydraulic ram? (We were destined to have a water-powered something!) The ram needs a piped flow into it, but a pipe, sluice, or aqueduct can carry the water to the ram site. Then, we ram part of the water up to storage and dump the rest back into the stream. Or a lower reservoir that's out-of-stream. Computing the flow rate of the stream and its duration, dividing for the ram's efficiency, we'd fill our 12,000-gallon closed reservoir in no time. Then, with a little smoke and mirrors, we worked in a turbine. When we've got all the water we need in the upper reservoir, the drive tube feeds the turbine, giving us power for as long as the stream lasts.

Selecting the device. Seasonal fluctuations, the application (mechanical or electrical energy), and the rate of flow will all help you select the turbine or wheel you'll build or buy, but number one on the list to look at is head. In other words, the head decides. Not yours. The water's head. Look at Fig. 4-10 and see for yourself. Considering the wheel sizes and other inherent limitations, how much effective head exists for the wheel or turbine is the deciding factor.

The application should get the next lookover. For mechanical work, the Pelton and Michell turbines are out (you need torque, not speed). Which one of three—undershot, Poncelet, or overshot—you use depends on the head, seasonal head variations, and the rate of flow. For electricity generation, the Michell turbine is the obvious first choice for the DIYer, and the Pelton if you're not a DIYer. If the head is under twenty feet but you've got an increasingly high rate of flow, you may use one of three or four wheel types, but prepare to do some extensive gearing; ratios as high as 300:1 are not uncommon! An alternate gearing method is illustrated in the Data Cubbyhole, Section I. The Michell turbine and overshot wheel are good with varying flow rates if your seasonal variations are severe.

Appearance is important in final selection. A Michell or Pelton in an enclosed shroud isn't much to show off, or look at. I'm not kidding. With all else equal, you may want to pick a wheel over a turbine just because it's picturesque. You wouldn't be the first or the last to do it. Nimble Michell turbine makers could use Plexiglas for the side panel of the shroud to display the "action." Watch out if you've high heads, though; higher pressures could blow out the relatively fragile Plexiglas.

If you find a manufactured wheel or turbine, congratulations. Contact the manufacturer and get the ratings, however, or your joy may be short-lived. A wheel or turbine that's mismatched to the water conditions—seasonal flow variations, head, or rate of flow—or to the application will perform unsatisfactorily. As well, what are you going to do if it breaks down and needs parts or, worse yet, replacing? Designing a system around a "used" part is always risky. Minimize it by checking its condition, the availability of parts, and its design ratings.

Installation siting. It may seem unnecessary to say this but: put your channel, dam, pipe, or turbine where it best fits. Every installation is different, but there's usually only one good way to put it all together. Here's some hints:

1. If you're not using all of the head that the site offers, choose the steepest part for the shortest run of pipe or channel, unless distance or some other factor forces you to select a less desirable spot.

2. Pick the shortest route between water pickup (by channel or pipe) and water delivery. Just because the stream wanders like a drunken cow through your place doesn't mean the pipe must.

3. Head may be measured between the high entry and the low exit of the stream on your property, but you may be able to take advantage of a lower point on the property if you can dump the waste water there. Obviously, the water won't be going back into the streambed. It's okay if you're going to fill a big pond or if there's another creekbed to which it can travel. Don't do this if you don't have the necessary water rights (see the section on legality later in this chapter) or if the runoff is going to cause erosion.

4. It's nice to put the waterwheel near the job. If you're just using the mechanical energy, site it close to the application. For distances of at least a quarter-mile, a taut-wire transmission line may work (see Sources and References section). For electricity production, put the powerhouse near the hub of activity. If it comes down to a choice of long pipe (bringing the water to you) or long wires (bringing the electricity from a remote powerhouse), opt for the wires. The price of piping water is usually much higher per foot than "piping" electricity.

5. If no other factor interferes, locate the dam, channel, or pipe for other water uses, tapping off to gravity-irrigate or to supply pressure for household needs.

6. Sometimes you wish that you had just a little longer section of stream on your property for the installation. Or you've got the ideal spot for the wheel but it's just over the property line. Just the opportunity you've needed to go meet a neighbor! If you're going to have enough power, you might consider starting up a little utility company, just the two of you. Think about it. No need to duplicate all of the equipment—it's just extra pipe. Or if there's not enough power to split up or you wish the minimum of contact with your fellow human being, offer something, make an agreement, shake on it, and put it in writing. Maybe even write it into the deed—each of yours. Nothing else works. You'd be a fool to plan an extensive installation on somebody's word alone. Even if he or she would honor it (and some would), you may have to deal with the next owner of the property, or heirs. If you can't get it in writing, but want to risk it, put the most expensive part of the installation—wheel/turbine, generator house, etc.—on your own property.

Legality. Doesn't seem like there'd be any problem with the law in plopping something down into the stream and getting some power, but there might be. It varies from place to place, so you'll need to check. It'd be a pity if you got all worked up reading this chapter and went out and laid some cash down for a pretty place with a ripsnorter of a stream and found out *after* the installation that you weren't permitted to do it.

There are two kinds of rights regarding streams—riparian and appropriative. The first doesn't cause much trouble because it's the law of "reasonable use." If water is running through your land, you've a right to use some of it for drinking, cooking, bathing, toilet, and watering the garden. If there's still some left over and nobody below you complains, you can use if for irrigation, filling a reservoir for the drought, etc. Damming it is another thing altogether, so check.

Appropriative rights are first-come-first-serve stuff and it's dog-eat-dog. Your rights are deeded, and you or they have "proven use" before someone else. It comes with the property (so many shares or acre-feet, etc.) or the guy ahead of you (upstream) uses it all. Just because the water's there doesn't mean that you can use it.

It can get sticky. You let somebody know you want to put in waterpower and all of a sudden someone's hollering and breathing down your neck. Is extracting energy from the water actually "using" it? Sure, you take it out of the stream but you can prove you put it all back; it leaves your property in the same condition. In fact, by borrowing it, it's cleaner (show 'em your trash gate!). And it's oxygenated for the fish! "Use," to the law, implies "use up." It's your best defense if anyone gets their dander up!

Water rights get treated a lot like building codes, which were made to *protect* the owner-builders and buyers, not *restrict* them as they do today. There's regular rebellion going on, and maybe you'll want to get involved in it, making your situation a test case. If you install something without checking first, you always run the risk of discovery later on, and whatever consequences it brings—your decision. My own belief is that a well-prepared presentation of what you intend to do, and a proper appearance and manner will win it through just about every time. Come prepared to answer questions (dazzle them with your knowledge) about the effects, benefits, and hardware and, in this time of energy shortages, it'd be hard to refuse. Good luck!

METHANE POWER

You've heard of pet rocks? Well, I've got a pet swamp. I keep it in a cage, too. Since swamps tend to escape between the bars on most cages, this one is special: it's in a steel tank—kind of hard to show him off, but we know he's there. Anytime I introduce someone to my pet swamp, I let 'em take a whiff of its delicate aroma. They all think it's a real gas!

And it is, too. What comes out of that little tube is biogas, and it's released anytime certain conditions are met as organic material decays. Our steel tank is a controlled environment. Without it, the material would decompose anyway, just as it's been doing on the planet Earth for a long time. And, as every faithful and diligent gardener knows, the compost pile works. When it's watered and stirred, folks can't help but notice the steamy, hot interior. This is known as aerobic (in the presence of air) decomposition, and it releases heat. Put the same batch of stuff into an anaerobic (in the absence of air) environment, like our steel tank, and gases are produced from the rotting vegetation or animal matter. One of these gases is methane (CH_4 to the chemistry bugs), and it's the same stuff that the utilities tap and sell to us as natural gases; their stuff is just a heck of a lot older than ours. Fortunately, the gas doesn't get any better with time, so ours is just as good.

What do we need in order to generate biogas? Other than the proper environmental conditions, very little. You could generate biogas in a coke bottle. Great fun, but hardly practical. How much organic material you'll need, how much gas it will produce, and what you can use the gas for is the subject of the following sections. Have at it.

BIOGAS

"Biogas" is the term used to describe the gaseous byproduct of the anaerobic digestion process. Many people think this is only methane, but it's not. Biogas is basically composed of five substances—CO_2 (carbon dioxide), H_2 (free hydrogen), H_2O (water), H_2S (hydrogen sulfide), and CH_4 (methane). The hydrogen and methane are burnable, but the quantity of free hydrogen is too small to go after. So, it's methane we want. The rest of the biogas components are vital to the continuation of life, but they're not useful for our purposes. In fact, they can be a downright nuisance. CO_2 is commonly used as a fire-extinguishing gas. Maybe you can see what a joke it is to try to cook with biogas—the methane is ignited and the carbon dioxide wants to put out the fire. Super. Having water in the gas is a bother, as well. Besides trying to drown our hypothetical fire, it consumes precious energy as it's turned to steam. Then there's a H_2S. No, it doesn't interfere with the fire itself. It's busy attacking the pipes; hydrogen sulfide is extremely corrosive to metals. And we're trying to cook with methane while the water is stealing heat, the carbon dioxide is playing fireman, and the hydrogen sulfide is determined to eat away the pipes carrying the methane to the burner!

Many biogas setups use filters to remove the effect of gases other than the methane whenever the application warrants it. Pass the gas through water, and the water vapor and CO_2 will be absorbed. Pass it over warmed iron filings, metal shavings, or old nails or through steel wool, and we leave the hydrogen sulfide behind. Leaving a little of each of the gases and a whole lot of methane.

I exaggerated a little. Just burning the gas won't always be that difficult. Engines that run directly off biogas seem to do well; if it's immediately combusted, the H_2S has little effect, if any. If the gas is to be stored (as a gas), or compressed into a liquid, precautions must be taken to protect the compressor, tank, and all fittings. Pumps with working nonmetal or specially coated metal parts must be used along with plastic pipe and nonmetal or painted metal tanks. Use a filter only if you're having difficulty. If you isolate the bothersome gas, the filter is simpler.

Filtering out H_2S is the tough one. Once your heated metal "getter" is coated with hydrogen sulfide, you have to replace it. Exposed metal is a must, and you won't always get it, since nature likes to coat these surfaces with her own idea of protection: rust. But maybe you'd like to leave the H_2S in there as an indication that your pilot lights or main burners have gone out—it has a horrible smell. Like rotten eggs. One of those chemistry experiments you can do to clear the lab in nothing flat. Or save it for the middle of a particularly boring lecture in another class. Everyone knows that propane smells, but what you may not know is that natural gas has no odor to begin with. The indelicate aroma is added! Otherwise you wouldn't know you were being gassed.

It's the same with methane. It has *no* odor. It's the hydrogen sulfide that normally accompanies methane which has the odor. And it immediately tells you when to replace the filter if you've been filtering it out. An olfactory indicator!

For reasons that will become evident later on, most uses of methane will be direct. You use it as a gas without compressing it. For that reason and others, methane systems are dangerous because they're begging for a spark—or a leak. It's not so much the methane getting out, as air (or the oxygen it contains) getting in. Buy a bottle of propane, connect it to a burner, and light it, and you'll have a nice fire to cook up dinner. The same goes for a bottle of methane. Buy a bottle of propane or methane *and* oxygen, ignite the burner, and you've just lit a fuse. It lasts a whole microsecond, too. If the concussion doesn't get you, the metal tank's fragments will. No effort must be spared in getting our biogas-producing digester airtight, and keeping it that way.

METHANE USES

It's a toss-up at this point whether to tell you about the uses for methane or how it's actually produced, but the former wins. I've already given you one example—cooking with gas—but there are others. How about fueling a water heater? You can also use it for refrigeration. The recreational vehicle craze brought many small refrigerators running on electricity or gas into the limelight. If they seem a bit expensive, try finding an old Servel unit; they were quite reliable and parts are still available (see Sources and References section).

The fittings on most stovetop, oven, and water-heater burners and refrigerator valves are for propane or natural gas, which have a higher energy yield than methane (per cubic foot of gas). In dealing with the same pressure (delivery from the tank or pipe to the burner), it should make sense that the fittings need to be enlarged, permitting a slightly higher percentage of methane gas to flow, burn, and produce heat of the same "quality." If you've got savvy for little things, rip the critter apart and drill out the jet hole, and you're on your way. If you're doing a doubletake on what I just said, get a refrigeration person to drill it out or order the replacement jets. Four things may prevent anyone from just drilling out the jet. One is inexperience, and we've already noted the alternative —have it done. Second is if you want to use a propane backup for the gas application, in which case you interchange jets (a two-minute job). Third, gas refrigerators don't like drilled-out jets. They balk and scream and waste gas and heat and make you wish you'd interchanged jets first off. And, fourth, you may need a different type of turbulator to go with the jet. What? The turbulator is the metal swizzle stick inside the housing that swirls the gas as it comes out. Don't ask me how, but that's supposed to help the air mixture—purity of flame and all that stuff. If I recall correctly, propane gets two grooves in its turbulator and natural gas (or methane) gets just one.

We've been talking about stationary uses. How about portable ones? Yes, methane can be used in cars (some of you may have heard about driving around the country on manure) and other places where "bottled" gas is ideal. Note that I said "can be." Not the same as "should be." There's a big pricetag on the portability of methane. Two things that add up to big bucks are the difficulty of compressing methane and, once again, hydrogen sulfide.

You'd have to reach back into the chapter on solar energy and the behavior of gases to fully appreciate this, but when you pressurize a gas, at some point it liquefies. And it stays a liquid, as long as you maintain the pressure. Opening the valve on a propane tank lowers the pressure and some of the liquid gas becomes a gas gas and that's what you burn. To liquefy propane takes a pressure in the vicinity of 300 psi, and it remains a liquid in any tank that can maintain around 230 psi. No problem there; a one-stage pump can handle it.

So, let's compress methane. Only, because it's a more basic gas, it's going to take a much higher pressure. As a matter of fact, over 2,000 psi. As with oxygen (for aqualungs), it takes a three-stage pump minimum. Expensive! And while the pump is laboring away at the methane, that darn hydrogen sulfide is eating chunks out of the metal parts. Better filter it or get a no-metal-exposed pump. Up goes the price tag.

For a large farm, a small community, or a city, it's a paying proposition. You only need one pump to compress all that methane for the many vehicles or uses. But for a small farm or a single household, liquid methane is out. Don't get too down in the face, though; once you figure out how much methane you'd need to feed the inefficiency of a car and how much organic matter that'd take, you'll wonder what ever made you think you would want to do it in the first place.

This is not to say that some people don't use methane for their vehicle(s) and have to liquefy it. On the contrary, you *can* compress it to around 200–300 psi, use a large tank (tow it behind you?), and drive off. Liquefying any gas crams the most gas into the least space, but you'd have enough this way to bop into town or take your gal (or guy) to the drive-in. A little awkward-looking and space-consuming, but it will work.

A final use for the methane is to use it to heat the digester to help make more methane gas. This sounds redundant or a little like perpetual motion, but it's next up and you'll soon understand why we'd do this.

Fig. 4-11. **Methane is difficult to compress; one might need to use up some passenger space to travel any distance.** *(Courtesy, James L. Ruhle & Assoc.)*

THE DIGESTER

Well, that makes about four times I've used the word "digester," and it's about time to define it. This is the special environment I've referred to which is conducive to producing methane. It's airtight and it has only trace amounts of oxygen in it. Gee, we're producing methane in a vacuum? No, the medium is water. Shove in the organic matter which is supposed to decay, and we've got our swamp. Because the process of decomposition involved bacteria which "break down" the matter and grow their own bodies umpteen different ways, we say that they "digest" the soup stock. A bacteria's restaurant: the digester.

Out of all the bacteria that are at work (or at lunch), there are some which produce a lot of gas. If we want methane, we must satisfy the needs and wants of these particular bacteria. When we go to a restaurant, we look for "atmosphere," and so do they. Catering to bacteria means paying careful attention to pH, temperature, and the S/L and C/N ratios. What do they mean?

pH. Straight out of the chemistry lab of our youth comes pH. The low side is acid, and the high is alkaline. It's a scale beginning with 1 and ending with 14. Right in the middle is 7 or "neutral." Good water is neutral. Methane production needs a pH environment between 7.5 and 8.5 to work best. When you first load the digester, it may be almost any pH. Once the air's removed, it may actually dip on the scale, becoming acid as the other type of bacteria goes to work. However, after several weeks, the pH should rise and the hungry methane bacteria will have their feast. They go from buffet to "buffered," eat merrily along, and we reap the methane.

If we load the digester too fast, introduce toxic materials, or influence any number of other factors, the pH can change. This hurts the methane bacteria and they stop eating. Ergo, less or no methane. A sick digester is usually capable of healing itself if given the time to do so. Stop shoveling stuff in and let it gain its strength, and the pH will rise and processes can resume. If the pH change is serious, remedies may be required. One may be the introduction of something alkaline, such as lime or ammonia water, to force it back in the correct direction. Rarely will a digester get too alkaline, but an acid may be added; since incoming slurry is acid, this is the recommended solution. Violent pH changes will kill the methane bacteria, "souring" the batch, and it may have to be dumped.

How do you know the pH? Remember that pink or purple paper that you dipped into solutions in chemistry lab? That's the stuff—litmus paper. It comes in different pH ranges and you test (following instructions) until you get the desired change of color or no change of color after wetted. Sticking it up the sewer pipe is like taking a baby's temperature. But you should have several places along the digester tank from which to take the readings; the pH is normally higher at the inlet end than the outlet point.

Temperature. Methane-producing bacteria are finicky about temperature, liking it about 95°F. Actually, they like the range between 85–105°. Below these values, methane bacteria get too cold and won't eat. Get it too cold and they die. Above this temperature, these bacteria don't like to eat, either. Another group of methane bacteria (brand Y) thrives around 130°F, but they're even more finicky, and it's a good thing they are too—it'd be very difficult to maintain that temperature for them. So, we stick with brand X—the 95°F group.

In the compost pile, heat is released. In the digester, gas is produced. But you can't have your heat and gas, too. In other words, we can expect no help from the digester soup in achieving this temperature or maintaining it for optimum methane production. On our own and with methods or sources we must devise, we must heat the soup for the eating. That's what I meant by using some of the methane to heat the digester. Of course, you may use as much methane as you produce in this method!

Another source of heat is the sun. Come to think of it, that's how the swamp gets its heat. With the correct number of collectors and adequate insulation around the tank, the contents can be heated to the upper limit (over 100°F) during the day and it shouldn't fall below 85°F at night.

If you can super-insulate your tank and live in an area that doesn't experience extremely cold temperatures, you may not need to heat the digester. A large portion (87%) of the slurry mixture is water; first heat it (by solar or other means) to well over 100°F, mix in the solids, stir to a cream consistency, and inject it into the digester. Sneaky but effective.

S/L ratio. Empirical research has determined a specific ratio of organic material to water for optimum digestion and methane production. By weight, it's about 87% water and 13% solids—a 6:1 ratio. A 55-gallon-drum-size digester, therefore, would have 350 pounds of water and 50 pounds of solids.

Gathering up 50 pounds of organic material and dumping it into the barrel is not going to work. If it hasn't been dried to remove the water it contains, there's actually very little "solid" to it. In fact, as much as 75% of it might be water. So you dry it, or let it dry. Until it's as dry as a cow pie that you can throw 75 feet. Even then it may still be as much as 25% water, but just figure it into your calculation.

Computations of the amount of gas released from a given weight of manure or other matter should take into account that not all of the "solid" is digestible. Chicken manure, for example, has a lot of grit (sand, dirt, etc.) in it; this is necessary to the digestive process in the bird. It doesn't matter how hungry any of the bacteria are, they won't eat it—it's inert stuff (I always leave the little pebbles in the bottom of my soup, too). But they may constitute as much as 25% of the weight, so compensate.

C/N ratio. The carbon-to-nitrogen (C/N) ratio is important in the digester; a 30:1 mixture optimizes methane production. That is, the bacteria prefer thirty carbons (the coffee) to one nitrogen (the cream). Anything else, and they go on strike. Each material that we might want to put in the digester has carbon and nitrogen, but in varying ratios. The trick is to find two or more different substances which, when combined, give a 30:1 C/N ratio. Easily said, but difficult to do. Tables which give approximated carbon and nitrogen content may not include the substance you wish to use, or yours may be older, dryer, or in some other way different. Too high or low a ratio will bog down (get it?) the digester. Unfortunately, there's no easy way to determine the carbon or nitrogen content of any given material. If you're getting methane, you've obviously got the right ratio. If you're not, the reason could be this or a dozen other factors for why your digester has taken a siesta.

Generally, we mix a high carbon with a high nitrogen material. Chicken manure is high in nitrogen. Straw is high in carbon. So, "barnyard straw" has both and would normally make a good mixture. Be careful, though. Selecting carbon materials which normally float on water is not going to work well. We need a mix to have an effective C/N ratio. The ideal slurry mix is fine, sticky (clings and can't readily separate), and nonlumpy. Hair, straw, and other floatables will form a most unwelcome scum at the top of the solution, well known for its ability to bring any digester operation to a halt. Scum is the main reason for providing access to the interior of the digester. It doesn't wash away—it must be broken up and removed. The consistency is similar to Rice Crispies squares.

TYPES OF DIGESTERS

I guess we'd better start talking about the equipment necessary to producing some practical methane. There are two types of units—one we fill a little bit each day, and one that gets a whole load at one shot. The first type is the kind used if you've just inherited your grandfather's turkey farm and you're wondering what to do about all those truckloads of droppings besides paying someone to cart them off. So you set up a *continuous feed* digester. Do it right, and you'll have set up a nifty natural gas supply more reliable than the stuff that comes via the utility pipe. Of course,

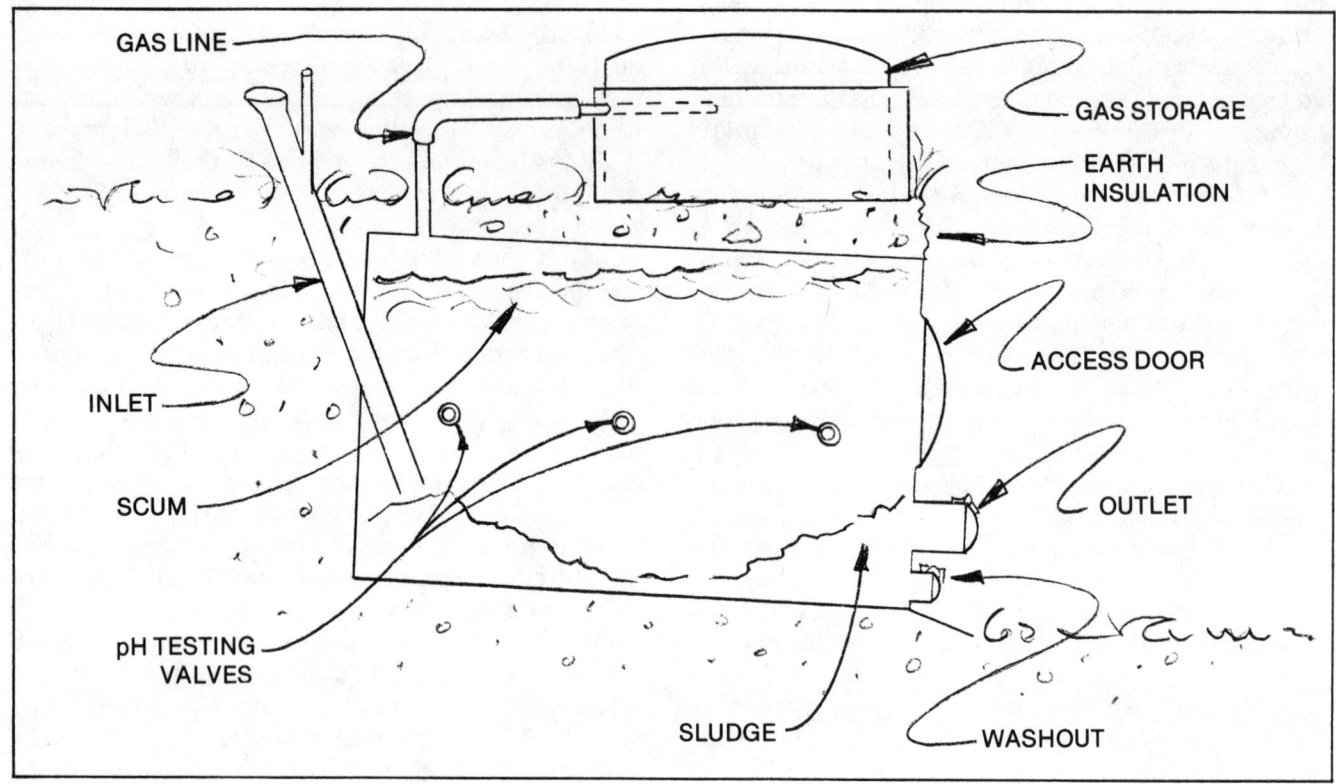

Fig. 4-12. A continuous-feed digester

you don't just shovel in manuare and get methane out; that's why you have the effluent pipe. The effluent is the processed organic material. Don't haul it away, however. Dry and bag it, and sell it as fertilizer, because it's the high-priced spread. Or spread it in a 1-2-inch layer over your own soil, and grow super-vegetables.

The *batch load* digester is usually the smaller variety of digester. It's the one used by the guy who'd been hauling away the turkey droppings before you came along. He gets a big truckload, hauls it to his little farm, mixes it with water in his homegrown digester, and seals it up. Whether or not he uses the methane that's produced isn't important; he's going for the end product for his fields: that rich fertilizer again.

DIGESTION OR METHANE?

We'd better get this settled right now. Are you building a digester or a methane generator? Sure, all digesters produce methane, and methane generators are digesters. Before you start making any plan, however, you should figure what you're after—methane or fertilizer. Of the two, the fertilizer is more beneficial, because you can grow things with it. Methane is transient, a by-product of the process. Digestion is better than using compost piles because it will deal with volume, it can be continuous-feed, and it will decompose the stuff faster. Even though you couldn't produce enough methane from the organic waste that's available to you, it might still be a wise decision to build a digester just to assure an excellent source of fertilizer. Plants of this size need not make any provisions for gas storage or use; it's simply vented away or, if the quantity is gassing your friends and neighbors, you can burn it. Might as well put the flame *under* the digester to help maintain the correct temperatures. Optimum methane production means an optimum decomposition rate; rarely can we expect simultaneous benefits such as these.

BASIC DIGESTER SYSTEMS

Fig. 4-12 illustrates the components of a basic continuous-feed digester system. It's composed of the main tank—its inlet, outlet, gas line, an access door, and odds and ends—and the gas reservoir.

The *tank* is the epicenter of activity and may be made from plastic or steel, for smaller units, and concrete for larger ones. This holds the slurry and it's affectionately named the digester. It must be airtight, a job in itself with all the openings it must have. It's heavily insulated or buried, to maintain optimum digestion and methane production. It should be heated. From underneath, if by a wood-or-methane-fueled fire (metal tank). Or from solar collectors, circulating a fluid. Do not put the tubes inside the tank; wrap them around the tank or along the sides between the insulation and tank wall. Access to the inside of the tank is essential for later removal of scum or grit. A horizontal tank will work better than a vertical one in this regard. The tank floor should slope gently downward to the outlet end.

The *inlet* pipe connects at a low point in the tank at the raised end; to prevent the muck from coming out when we want to put some new brew in, we use a pipe that slopes upward at an angle until its upper end is above the top of the tank. A long ram will help move the slurry (mixture of water and solids) into the tank. A straight inlet tube allows us to clear it if it becomes clogged or blocked, so no 90° elbows, please.

The *outlet* pipe gives us the effluent (sludge and supernatant)—decomposed vegetable and animal matter. It's not quite at the lowest point of the tank's end—that's reserved for the sand and grit outlet—but it's close. Obviously, with a full tank, we empty some of the effluent out before we try to put some slurry in. Wouldn't want the soup to overflow!

The *gas* line leaves the tank at the highest point; methane, being light, rises to the top of the tank. As more is produced, a small amount of pressure is formed, and the gas will move out the pipe. If no use of the gas is intended, it is simply vented to the atmosphere. To prevent a deadly mixture of gas and oxygen *inside* the tank, the gas line should terminate in a jar, bucket, or barrel of water, forcing the gas to bubble up through the water before it is released. In the event that the methane production ceases or a slight suction (vacuum) occurs in the digester (drawing off effluent), only water can be sucked down the gas line. Think of it as a one-way valve—letting methane out but no air in.

Other openings may be needed in the tank. A sand or grit trapdoor should be built at the lowest point of the tank. You'll appreciate this feature if you draw the short straw and get to go "into" the clogged-up tank after it's been drained. Even if it has aired for a few days, it won't be one of your more pleasant experiences. It will be mercifully short if you can hose down the interior rather than shovel it out. A few other valves should be located at select points for testing the pH of the brew. Locate a few large plate openings in the top for scum removal; in this way, the tank won't need to be emptied just because you must remove the built-up scum.

If you use the methane gas produced, it must be stored in some kind of a *gas reservoir*. The main design feature is to keep it from mixing with any oxygen. This is assured in the inverted drum-within-a-drum method (see Fig. 4-13). With a little water added for a check valve, the inverted drum will fill with the gas and lift up, announcing that it's full. You'll need several. A rock or two on the tank will "pressurize" the gas to some extent for use; experiment to find the right weight your burners like.

Fig. 4-14 illustrates a simple batch-load digester. If it's small enough, the inlet can become the outlet by merely tipping over the unit. Larger ones will re-

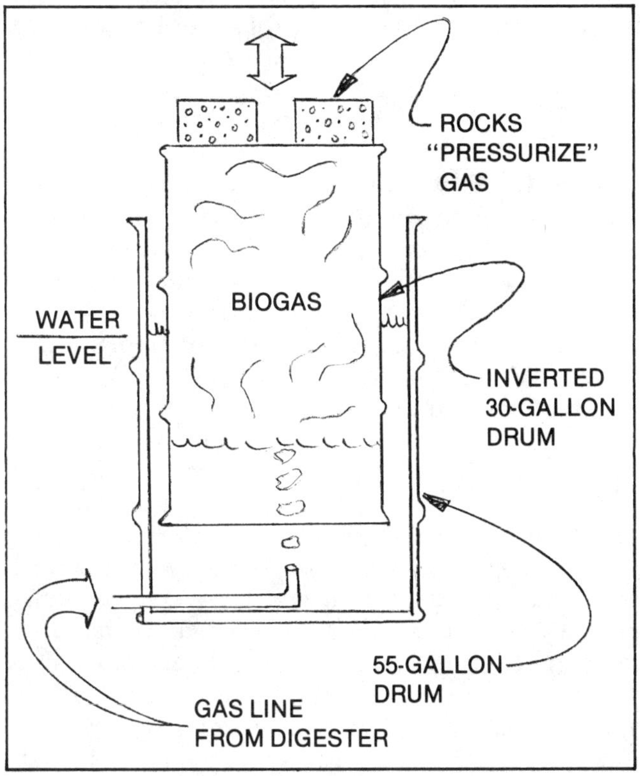

Fig. 4-13. A simple biogas reservoir

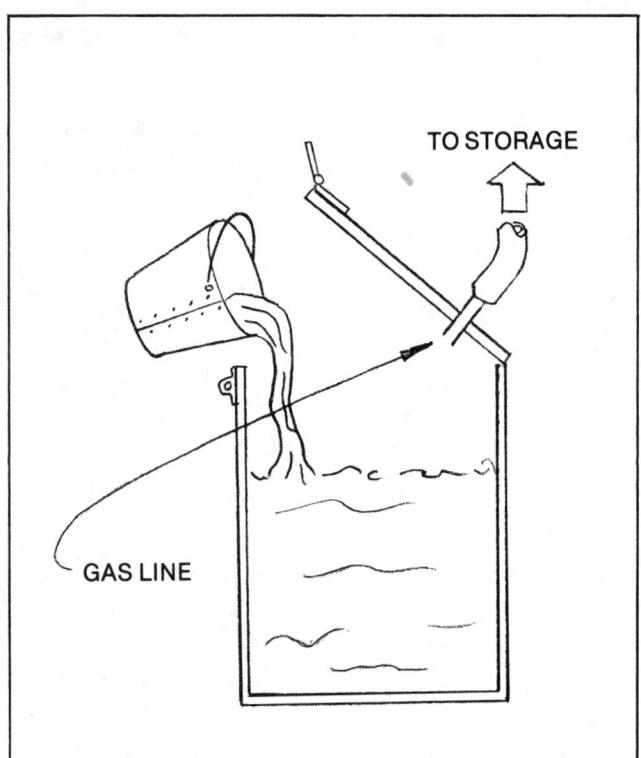

Fig. 4-14. A batch-load digester

quire separate inlet and outlet openings. Put the inlet where you'd put the scum access door in the continuous-feed unit—in the very top; we don't have to worry about air contamination during loading. Once the batch is in, seal the inlet and wait until it's all done; just come back in a month or so.

Sizing and yields. Biogas has two-thirds the Btu value of natural gas. The lower value may be attributed to the relatively low percentage of methane in biogas (about 70%) and the high value in natural gas—about 90%. The other factor is the high yield (30%) in biogas of fire-quenching carbon dioxide, which is totally absent from natural gas. There are approximately 650 Btu's of heat in each cubic foot of biogas.

The best gas yield we can expect from a digester/methane generator is 5 cubic feet of methane from each pound (dryweight) of sewage and farm waste. Figure a conservative 50% in your design calculations on "need," and you'll be pleasantly surprised anytime it exceeds this.

Digesters detain the slurry/sludge for about thirty days. If a continuous-feed digester system is planned, each new load should not exceed .25 pound (dry weight of solids) per cubic foot of tank or .03 pound per gallon of tank. Size the tank according to the load you have, or size the load for the capacity of the tank you found. Since the S/L ratio is 6:1 (87% water and 13% solids), you have all the necessary info to make the estimates either way. Let's take two examples, in case I lost you.

A 400-gallon system. You've got a 400-gallon tank. Leave 10% of the capacity for gas accumulation, and that means 360 gallons of working brew. Figure thirty days to complete one cycle of slurry-to-effluent; that means you load in 12 gallons of fresh slurry per day. Twelve gallons of slurry, at approximately 8 lbs. per gallon, is 96 lbs. of input; round it off to an even 100 for easier calculations. With the solids composing 13% of the slurry mixture, we need 13 lbs. (dry weight) of solids per day. At 5 ft^3 of gas per pound of material, that's 65 ft^3 of gas per day and, at 650 Btu's per ft^3, that's well over 42,000 Btu's of available heat. Even if we consider getting only one-half of this value, that's a lot of heat.

Computing it by a different method, we've got 50 ft^3 of tank (400 gallons divided by 8 gallons per ft^3), and leaving 10% for the gas, that's 45 ft^3. Meaning we load 1.5 ft^3 of slurry per day or, at .25 lbs. (dry weight) of solids per ft^3 of tank, we have 11.25 lbs. (.25 multiplied by 45 ft^3) of solids to add each day. That's pretty close to the 13 lbs. computed by the gallon method (above).

A 50-gallon system. Your digester is a 55-gallon drum. Lop off 10% for gas and we've got 50 gallons of soup. Feed it every day and allow thirty days for the cycle and we need to add 1.6 gallons of slurry per day. That's 12.8 lbs. of mix and, at 13%, there will be 1.6 lbs. of solids to add. That's 8 ft^3 of gas or 5,200 Btu's per day maximum gas production.

If you work in cubic feet, a 55-gallon drum has 6.8 ft^3 of capacity. Leave room for the gas and we've got 6 ft^3 of broth. At .25 lbs. per ft^3 of tank, we need to add 1.5 lbs. (dry weight) of solids per day.

Operation. A starter must be used when "charging" a digester for the first time—sort of like sourdough starter. This is stuff where we know we've got virile bacteria of both kinds. A sewage-treatment plant can give you some; bring your own jar. If you've batch-loading or starting a continuous-feed unit with a high initial load, bring a 55-gallon drum for the starter.

Don't hold your breath for the first bit of methane; we're talking about three to five weeks. You might start getting gas any day, but that's mostly CO_2. You're running it through that little check valve, right? Each day you can go out and hold a flame to the bubbles coming out the jar, bucket, or barrel of water and you'll inevitably smile when you get your first ignition. Once you've got a good positive pressure and can support a flame from the end of the gas tube, open the line into the drum storage unit and go treat a few friends to a different kind of brew.

Keep records of pH tests, temperature, loads (batch or every day), remedial actions, etc. Get sloppy here, and that's the way your digester will work. It's simply amazing how meticulous records can help nurse the culture. It's a living thing on the other side of that tank wall (I hope it doesn't give you nightmares), and its needs are definite.

CODES

Out on the farm, the septic tank or cesspool is the establishment answer to waste disposal. Its origin is buried in legislation, politics, and rampant ignorance. Your chances of convincing a sanitation inspector of a digester substitute for human waste is zilch point zero. It may, however, be approved for animal waste or vegetable matter. Creates a dilemma, doesn't it? Actually, you can't blame the system; human waste is chuck full of nasty germs and parasites which will do you in quickly if you don't kill them. So, you have three ways to go. Bone up on everything about digesters so you can really "talk shop" with the inspector. If you can convince him you are well-informed and deadly earnest, you may get the go-ahead on a provisional or experimental basis. Or check the codes for your county; you may be living in one of the rare places where a digester for human waste can be approved. Or, build the digester for other types of waste, and use a code-approved system for human waste. There are other options, most of them illegal,

Fig. 4-15. A community-scale methane production setup in India *(Courtesy James L. Ruhle & Assoc.)*

and all guaranteed to get you in trouble if you're found out.

DISCLAIMER

There is insufficient information in this section on how to actually build a methane generator or digester. Even if you don't know it, I do. It's the truth. You must arm yourself with more information of a practical nature. The first two works listed in the Sources and References section for methane info are excellent, and both should be on your bookshelf before you start any construction. Controversy still rages on a number of points, and you will sometimes find opposing viewpoints in these two books alone, but read and heed what they agree on, and decide on the rest. I hope that this section has given you enough information on methane so that you can decide whether there is any hope of applying it in your own life. If I reached that point for you, I'm happy; that was the goal. Happy digesting—urine the methane generation!

5. Integration

Even a simple excursion into the world of alternative energy will reveal the possibilities and paths that abound for anyone interested in making the appropriate changes. Even to the casual observer, patterns become obvious. The more you know, the less you pay. The effort spent in conservation of energy nets a greater payoff than that devoted to collecting it in the first place. Finally, there's the sobering effect of TANSTAAFL: There ain't no such thing as a free lunch. Everything exacts its toll—whether it be money, time, energy, knowledge, or change.

Any source of energy may fulfill the requirements of several applications, but it usually only does one or two things effectively. If we need heat, we can call upon solar or wood energy. If it's electricity we're after, we use water or wind energy. Electricity from the sun or heat from the wind, however, is (at this point) asking too much of a slim pocketbook and any regard for conversion efficiencies. Know what works best where and how well.

Too often I see people trying to confront the problems they encounter with brute force. That is, if something consumes more power than how much is being made, make more power. Bust through there! Or skirt around it somehow. Or ignore it. And, when these tactics fail, sit down and bang the old head against the wall in frustration. Or do without it. It's a tendency we must fight. Surely, if you or I walk into a blind alley, we don't do any of these things. Rather, we turn around, retrace our steps, and seek another route.

Integration is a graceful blend of old ideas and new applications—knowledge and experience applied to tangible needs. If the conclusions are unacceptable, re-examine the premises and assumptions. Tackle each problem at a system level. Here are a couple of examples.

REFRIGERATION

Refrigeration is one of the biggest headaches to the modest wind-electric system, consuming the largest number of kilowatt-hours monthly—assuming you agree that modest means *no* electric ranges, electric water heaters, dishwashers, or air conditioners. Admittedly, the refrigerator doesn't consume an awful lot of power, but it adds up to a lot of energy considering how often it must run to keep its cool. What to do?

The first step in dealing with this dilemma is to divide it into component shortcomings. First, it's observed that the refrigerator turns on frequently during the day even when the door is *not* being opened and food inserted or removed—this is an insulation problem. Second, when the door is opened, cold air "falls" out of the interior—a design problem. Third, the refrigerator's compressor seems to labor extra long and we note that it is backed against the wall—with insufficient airspace about the heat-dissipating coils, we have a siting problem. Fourth, we're looking at a 12-volt DC wind-electric system and wondering how we're going to match it to a refrigerator that's rated for 110-volt AC, 60 cycle—an engineering problem. Fifth, food and drink are sometimes placed in the refrigerator at above-room temperatures—an operator problem. Frequent openings of the door also occur whereas a little forethought would ensure the removal of everything needed in one fell swoop—another operator problem.

Once identified, the solutions trickle in. Pulling the refrigerator out from the wall partially answers problem three but detaching the motor/compressor and heat coils and locating them away from the refrigerator altogether (outside the building in the shade?) wholly eliminates the third issue and partially answers the first. In the meanwhile, it may occur to you to buy an inverter (an electronic device that transforms 12-volt DC into 110-volt AC, 60 cycles) to mate wind-machine and refrigerator, but this would be a costly and inefficient mistake. Better yet, give the motor compressor to any friend in need of a small boat anchor and find a compressor that operates with a bolt-on motor. Next, find yourself a 12-volt DC motor capable of handling the load, and mount it. *Voilà*, you have a 12-volt refrigerator! Hire a refrigerator person to re-connect the motor/compressor with the stripped refrigerator and "recharge" it with refrigerant.

Now, lay the refrigerator on its back so the door opens like a chest-type freezer; this solves problem two. You'll have to build different kinds of shelves to accommodate the food and occasionally dig for things, but it's worth it. Next, add extra insulation around the refrigerator (bottom, back, sides, front, and top); rigid foam or fiberglass will do fine. Add some paneling or woodwork to dress up your handiwork. Problem number one is gone. Educate users of the refrigerator to think before opening the door, and to always allow foods to cool to room temperature before adding them to the refrigerator's workload. Exit problems five and six.

Is it worth it? With the exception of frost-free refrigerators, these modifications will reduce the monthly power consumption by 75% of the original figure or better. That's a good deal no matter how you slice it!

COOKING

Next to water heating, cooking is the biggest consumer of gas or electricity. Wood is a natural alternative, but slaving over a hot cookstove indoors during the summer months won't get anybody's vote. Weather permitting, a good alternative is the solar oven. Build it big enough to handle bread, vegetables, and stew and soup-size pans and it will do it all. Slowly but surely. It will, however, need to be turned (oriented). That means mounting it on a pivoting table or scratching some lines on the table to indicate how far it moves each time. Carry a small egg timer that will give you a crisp "Ding" when the twenty minutes between shifts is up.

It's long been claimed that foods which simmer and cook slowly have better flavor. True or not, at least there's no worry about burning the food if you forget about it; without your watchful effort, the oven stops cooking after twenty minutes, as the sun moves on. If this is more of an inconvenience than a blessing, automate the tracking and go climb a rock.

Foods are cooked differently in a solar oven. Since a few failures are likely to turn you off entirely to the concept, check out some of the nice recipes tried and proven over twenty years from Dan and Beth Halacy in their new book (see Sources and References section).

A STANDBY GENERATOR

Homebuilt or store-bought, a standby generator is a good starting place when making the transition to alternative energy sources. Even if it is gasoline-fueled, it's a necessary first step in medium wind areas if you're planning to install a windmachine eventually. If you wish to remain with utility power, it helps until you're tied in, serves as backup in a blackout, and may be the only thing to power your submersible well-pump when that raging forest fire wipes out the utility lines in its relentless march toward your land.

Simplicity is a wonderful attribute of the standby generator, but silence is essential. A 30-gallon barrel buried in the ground makes an excellent muffler. Or, better still, bury the beast itself in the ground (in a packing crate or container, of course).

A rather interesting arrangement is to start off with a water-cooled engine. Dig a hole, insert the container for the engine, and then the engine. Dig a trench between this unit and the house. Install electric wires, waterpipes, and air ducting and insulate it to the teeth. Close the trench and bury the engine compartment, leaving an access tunnel and trapdoor. Connect everything up. What do we have?

First thing in the morning, it's an easy matter to start the engine via the remote control panel in the house; since you'll hear none of the engine noise, you must have indicators to tell you it's going. Once they're positive, you get to go start breakfast. The engine's alternator is pumping power to the battery bank for a day's worth of electricity. As the engine warms, the water coolant circulates to the house where you've enclosed the radiator in a tank; in half an hour, you'll have 30 gallons of hot water for your shower. The engine's fan sucks air from the house via the inlet air duct and blows air, warmed by passing it over the engine, back into the house, taking the bite out of the previous night's cooling. You're still in the kitchen, monitoring the cooking of breakfast in the special-built, engine exhaust-heated oven. You get a hot shower, a warm house, breakfast, and a charged set of batteries and the engine consumes a quart of gasoline!

FINAL WORDS

A lot of what's been talked about here are old ideas incorporated into new applications. Think it out, do whatever portion you can yourself, and farm out the rest. For these and other needs, avoid complexity and ornamentation. Remember the KISS principle: Keep it simple, stupid. It'll keep you smiling. Good luck!

Appendix: Data Cubbyhole

I like continuity. When I'm zinging along in my writing, and come to a place where a little more explanation would be nice, or an example would do the job, I find myself reluctant to work it out just then. I make a mental note, tucking it into my "cubbyhole," and go on. It's kind of like sticking your peas into your pant cuffs when the dessert arrives. Sooner or later, though, you gotta drag 'em out. And then the only place left is the back of the book (for the examples, not the peas). So here they are.

SECTION A: SIZING SOLAR ENERGY STORAGE

The toughest job in using solar energy is this calculation. A rule of thumb would be most convenient, but seems very elusive. If you tried to make a formula that would take into account all of the factors involved in calculating collector size, storage capacity, application, and geographic position of the final installation, it would scare you to death! We are talking about five factors that involve heat or its absence: solar influx, storage, application (space heating?), auxiliary (backup heating), and ambient (outside) temperature. Here goes.

Solar heat values must be calculated in terms of the Btu's one could expect to collect, per square foot of collector, each day—minimums and maximums. Take a reading each day of the year and hope that next year is identical. Or, take weekly or monthly average days for each of the fifty-two weeks or twelve months. This information may be available from a local climatological station. Determine the number of cloudy days and the longest number of days the system should supply heat. Figure total collector area.

Storage involves picking the medium (rock, water, or salts), figuring the density of the material, its specific heat, the volume (gallons or pounds), and the difference in temperature between the lowest desirable temperature in the house (65°F?) and the highest it can efficiently store. Compute heat loss from storage (through its own insulation). Multiply Btu's of heat required per day times days of desired storage capacity.

Space is intrinsic to the house and your idea of comfort. How much volume of space in the structure needs heating? It's normally referenced to square feet of floor and assumes an 8-foot ceiling; adjust accordingly. Determine heat loss of house—through walls, floor, and roof.

Backup is any assistance that must be rendered during long spells of cold. Even if you assume that there will never be any failure of any part of the solar collection system, backup should be able to assist during freak storms. A 20% capability should be the minimum.

Ambient is the environment that is trying to steal the heat. This involves your specific locality and the effects of latitude, altitude, shading, and wind or heat loss.

When you've completed the study on how to make all of these computations, then compute for cooling. A system which is designed to heat a space will not necessarily cool that space during the summer. See Sources and References section for material explaining these calculations.

I'm not trying to put you on; it's an involved process. Years ago, I got tired of all the calculations and worked up my own rule of thumb. You need 2–10 gallons of medium and storage per square foot of collector for air systems. A 1–2-gallon capacity for water-heating systems per square foot of collector was adequate. This sytem works every time, because I always leave room for more collectors or more storage, or less of each if it's too much. When lacking a rule of thumb, make one up and move on.

SECTION B: ANGLE OF DECLINATION CALCULATIONS

Calculations for the angle of declination in winter and summer (the two extremes) will be much easier to make if you realize three very important facts. One is that the sun's rays are essentially parallel; keep that in mind as you look at the drawings. Second, latitude is measured from the center of the earth. Figuratively, that is, since no one has ever been there. The equator is 0° and the north pole is 90°. Each 1° or 10° of latitude is the measure of an arc corresponding to the angle. Latitude 52° north is a line that extends all the way around the globe, and it's that circle on the sphere that's exactly 52° *above* the equator line; there is a 52° south as well, which is below the equator. And third, the earth is "tilted" on its axis with respect to the sun, and that angle is 23° (see Fig. 2-35).

Let's look at the extreme angles of declination for three locations. First we visit mountainous Quito, Ecuador, on the equator (Fig. 6-1). The latitude is 0°. On December 22, the sun is in the southern sky at a declination angle of 23°. On June 22, the sun would be at the same angle, 23°, in the northern sky. On the equator, then, the coolest times of the year are when the sun is at either of these extremes; imagine having two summers and two winters!

The job opportunities aren't that great at the equator, so you visit a friend in Mazatlán, Mexico, which has a latitude of 23° north. On December 22, the sun will be directly overhead (Fig. 6-2), giving an angle of declination of 0°. On June 22, the sun will be in the northern sky at an angle of declination equal to 46°. This should make sense in light of the fact that the "swing" in the sun will be a total of 46° and we started from 0°.

You get an answer to a letter of inquiry for a job in Monterey, California, and find, when you go to design your collector system, that the latitude is 37° north. Therefore, on June 22, it is a little south of directly overhead (Fig. 6-3) by an angle equal to the *difference* between 37° north and 23° north (where it was 0°, remember?). Therefore, this is 14°. Computing the angle of declination for December 22 is easy; just add 46° to 14° and you arrive at 60°.

A fixed collector set at latitude 37° north should be set at a tilt angle to optimize the winter sun's energy (Fig. 6-4). The rule of thumb for collector angles is: add 15° to the latitude. In this case, we'd have 15° plus 37°, which is 52°. Makes sense. At latitude 23° north, we'd have a collector angle, using the rule of thumb, of 15° plus 23°, or 38°. At the equator, you might just point it straight up or design an adjustable collector angle to account for the two "winters."

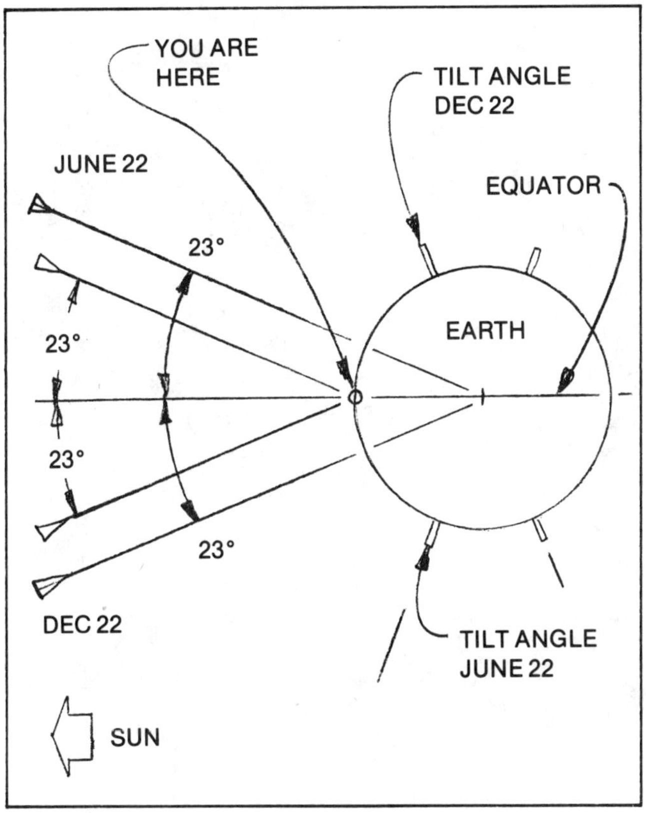

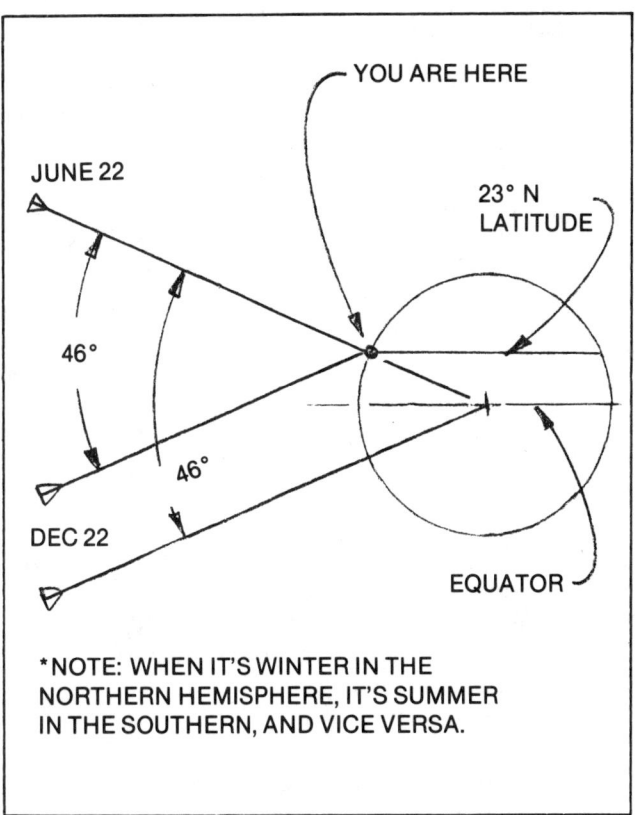

Fig. 6-1. The declination angles at the Equator

Fig. 6-2. The declination angles at 23° North

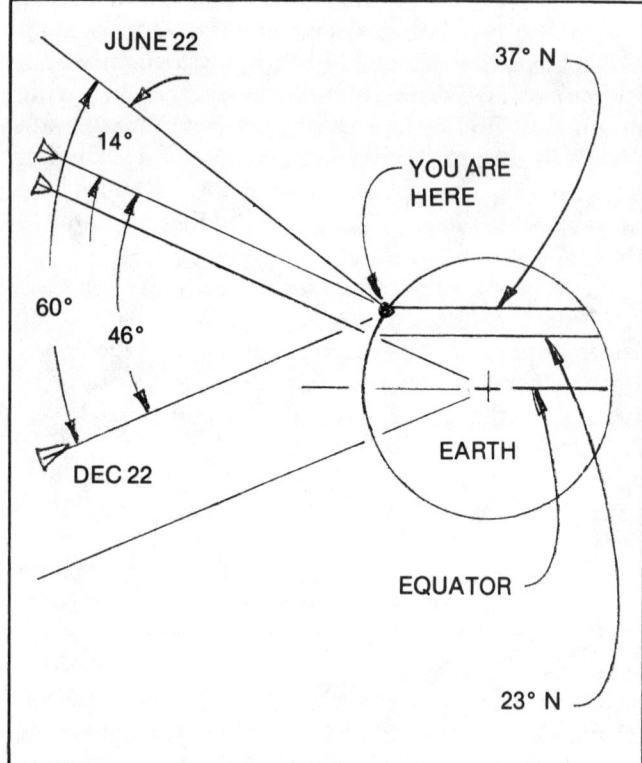

Fig. 6-3. The declination angles at 37° North

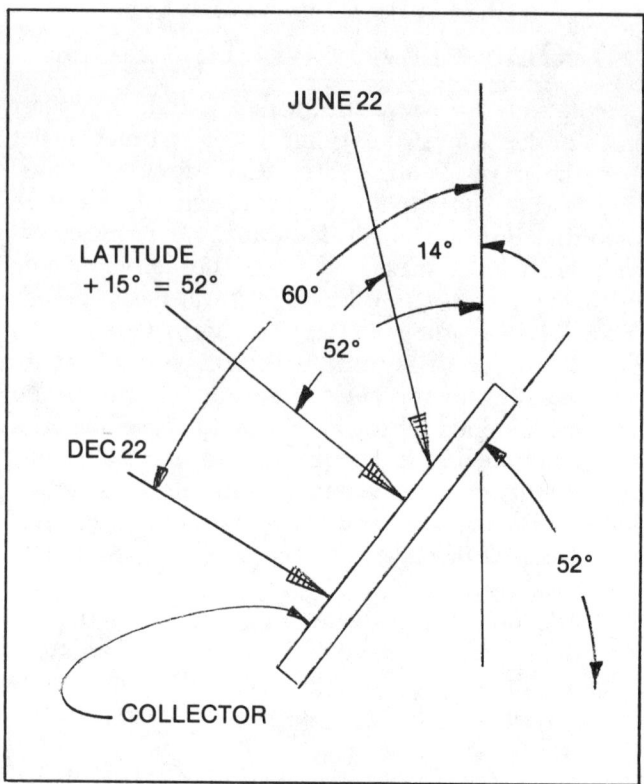

Fig. 6-4. Optimizing the declination angle for a fixed-position collector at latitude 37° North

SECTION C: AEROTURBINE EFFICIENCY AND THE WINDPOWER FORMULA

The base figure in the power formula (Formula 2, under the head "How Much Energy?") is 0.0030685, but we may combine it with E, or the aeroturbine's efficiency, once we've chosen the aeroturbine type and estimated its efficiency. That further simplifies the formula for frequent use. I've prepared a chart (Fig. 6-6) from my calculations to save you the work. To use it, estimate your aeroturbine's efficiency (from the next section, C-2), find the value in the "% efficiency" column, read across, and that's your new "multiplier." A 35% efficient aeroturbine, then, would have a multiplier of .0011 and the new formula would now read:

(1) $P = .001 \, AV^3$

Once you've computed the area of your aeroturbine, multiply A by the multiplier and the formula will be reduced further. If, for this example, we calculated the completed aeroturbine at 140 ft², our formula would now read:

(2) $P = .0154 \, V^3$

and you'd just pop in the velocity, cube it, and hit it with the multiplier to find power output.

Fig. 6-5. Earthmind's water-heating, parabolic tray collector

APPENDIX 129

% Efficiency	Multiplier	% Efficiency	Multiplier
5	.00016	35	.0011
10	.00031	40	.0012
15	.00046	45	.0014
20	.00061	50	.0015
25	.00076	55	.0017
30	.0009	60	.0018

Fig. 6-6. Combining aeroturbine efficiency with the base multiplier

SECTION D: AEROTURBINE EFFICIENCIES

The following percentages represent the upper limit of efficiency of these types of aeroturbines. Materials of greater weight, poor design detail, and lousy construction techniques will all work to lower the aeroturbine's final efficiency in extracting energy from the wind and transferring it to useful work.

Propeller . . . . . 2-blade 50%
                 3-blade 45%
                 4-blade 40%
Windsail 30%
Turbine 45%
Dynamo 45%
Savonius rotor (S-rotor) 15%
Darrieus rotor (D-rotor) 35%
Hybrid (S-rotor and D-rotor) 25%

Note: These figures are here to help you complete the information necessary to make power-formula calculations. Selecting an aeroturbine type based upon a list of efficiencies would be a mistake; don't make it. Evaluate all of the factors involved before selecting any type of aeroturbine.

SECTION E: WINDPOWER FORMULA CALCULATIONS

I know. You've been chomping at the bit to tear into that power formula. Here's your opportunity. Use the base formula (P = .0030685) or include the efficiency figures by using the chart in Fig. 6-6 for our examples.

Example 1. A 40%-efficient, 3-blade propeller unit 12 feet in diameter in a 23-mph wind will produce how much power? First, we find the swept area. A 12-foot blade has a 6-foot radius. Square it and you get 36. Multiply that by π, or 3.14, to get the area, which is 113 ft². Therefore, our formula is:

(3) $P = (.0030685) AV^3 E$
        $P = (.0030685)(113)(23)^3(.40)$
          $= (.0030685)(113)(12,167)(.40)$
          $= 1687.5$ watts or 1.7 kilowatts

Example 2. A 15%-efficient Savonius rotor 12 feet high and 8 feet wide is exposed to an 18-mph wind. How much power should it produce? The swept area is 12 multiplied by 8, or 96 ft². Let's use the table in Fig. 6-6 and, at 15% efficiency, we find a multiplier of .00046 and our formula is:

(4) $P = (.00046)(96)(18)^3$
         $= (.00046)(97)(5832)$
         $= 257$ watts

SECTION F: ESTIMATING WIND AT VARYING HEIGHTS ABOVE GROUND LEVEL

Over a flat plain, the wind velocity (V) at a height (H) may be estimated from a reading (Vo) taken at a height (Ho) from:

(5) $V = Vo (H/Ho)^{1/5}$

This works for windspeeds (Vo) between 5 and 35 mph; above this value, we use 1/7th instead of the 1/5th. Do you recall what the 1/5 figure is when placed as it is? That means you take the 5th root of what's inside the parenthesis. Boy! I bet you don't have one of those on your calculator! And unless you have some weird tables, you have to guess. With a calculator, it's not so bad. Enter a hypothetical figure and multiply it by itself five times. When you get a number that results in something close to the figure inside the parenthesis, that's the fifth root.

Let's take an example. You measure the windspeed with a hand-held unit and it registers 12 mph at a height of 6 feet off the ground. You want to know the wind's speed at 30 feet. We have:

(6) $V = (12)(30/6)^{1/5}$
         $= (12)(5)^{1/5}$
         $= (12)(1.38)$
         $= 16.56$ mph at 30 feet

Where did I get the 1.38? I did exactly what I told you to do—estimated! And got so wrapped up in it that I figured out the fifth root of a whole bunch of values, which I put into a chart (Fig. 6-7). As long as you take all of your readings at 6 feet off the ground, the chart applies directly.

Now who lives on a flat plain? No rolling hills, no trees, houses, obstructions? This formula, nevertheless, is intended for a flat plain, meaning it's not designed for any other place. However, I ignore the qualifier, figuring that even an inaccurate formula is better than *no* formula. As well, after years of inexperience, I've found that the tower is rarely built from a formula. Rather, you put up a tower that's as high as you can afford and as safe as you can raise, guy, climb, and work on. Good luck.

If (H/H_0) is	For a H_0 of 6 ft, H is (in feet)	5th Root of H/H_0 is
2	12	1.15
3	18	1.20
4	24	1.32
5	30	1.38
6	36	1.43
7	42	1.48
8	48	1.52
9	54	1.55
10	60	1.59
11	66	1.62
12	72	1.65
13	78	1.67
14	84	1.70
15	90	1.72

Fig. 6–7. Windspeeds off the ground

SECTION G: FLOW RATE CALCULATION METHODS FOR SMALL STREAMS

There are three ways to measure the rate of flow for a stream: capture, flow/section, and weir. Which one you use is determined by the approximate flow rate of the stream, how accurate you want the calculation to be, and how lazy you are.

Capture. For small streams or creeks that won't fill a 55-gallon drum in one minute, the capture method might work. *If* you can find a spot where it drops through a narrow spot. What you're going to do is take a container of known capacity (1 gallon, 5 gallons, etc.), place it so that *all* of the water in the stream runs into it, and time it until full. Since flow rate implies quantity and time, timing the filling of a container of known size gives us a rate. Then you convert this to whatever units you want, and you've just measured the flow rate of your stream—at the time you measured it, of course.

Let's do an example. Your miniature Mississippi fills a 5-gallon gas can in 15 seconds. That can be reduced to gallons per second or gallons per minute. We'd have 1 gallon in 3 seconds, .33 gallon per second, or 20 gallons per minute. Gallons per anything isn't all that useful to use, so we need to convert this to cubic feet per minute or cubic feet per second. Sound difficult? Not really. There are approximately 8 gallons in a cubic foot of water. So, to get our result into cubic feet per second, we divide .33 gallons per second by 8 gallons/ft^3, and .04 cubic feet per second is the answer. That wasn't so tough, was it? That doesn't sound as much as one minute's flow—which would be 60 seconds multiplied by .04 cubic feet per second, or 2.4 cubic feet per minute—but it's identical. Incidentally, dump that water out of the 5-gallon can, lest you think it's gas and put it (in an emergency) in your car or truck!

Fig. 6–8. Jim Davis and the author gain a new perspective on things

The capture method should be repeated several times until you get consecutive readings that are the same; there's a tendency to fudge, but don't. The trouble is, however, that the stream may meander instead of drop through too narrow a space for you to capture all of the water in this test. Get it all or you're just guessing how much doesn't go in. If you can't even temporarily dam the stream to get the flow through a pipe sticking out of the dam, abandon this method and try for the remaining two. Oh . . . don't be discouraged by the small amounts of water a successful use of this method may give you; flow rate isn't the only criterion for waterpower.

Flow/Section. As previously mentioned, flow rate can be found from finding the water velocity and the stream's cross-sectional area. While this is not as accurate as the other two methods of figuring the flow rate, it may be the only alternative in some situations, and it can be less work than the weir method if the capture method is impossible.

Streambeds are highly uncooperative when finding the cross-sectional area is important. If we could momentarily freeze the stream's flow solid, slice into it, and lift out the section, we'd have exactly what we want. But there's another way, admittedly less accurate, but less consumptive of our latent powers (Fig. 6–9). Mark a line or length of wood into equal sections (we'll call this the rod); the more the merrier, but there are limits. Set this on the perpendicular across the stream, bank to bank, and get it reasonably level (break out the bubble level). Make a plumb bob. A what? You know, a heavy weight on a string to indicate "down" in our gravitational field.

APPENDIX 131

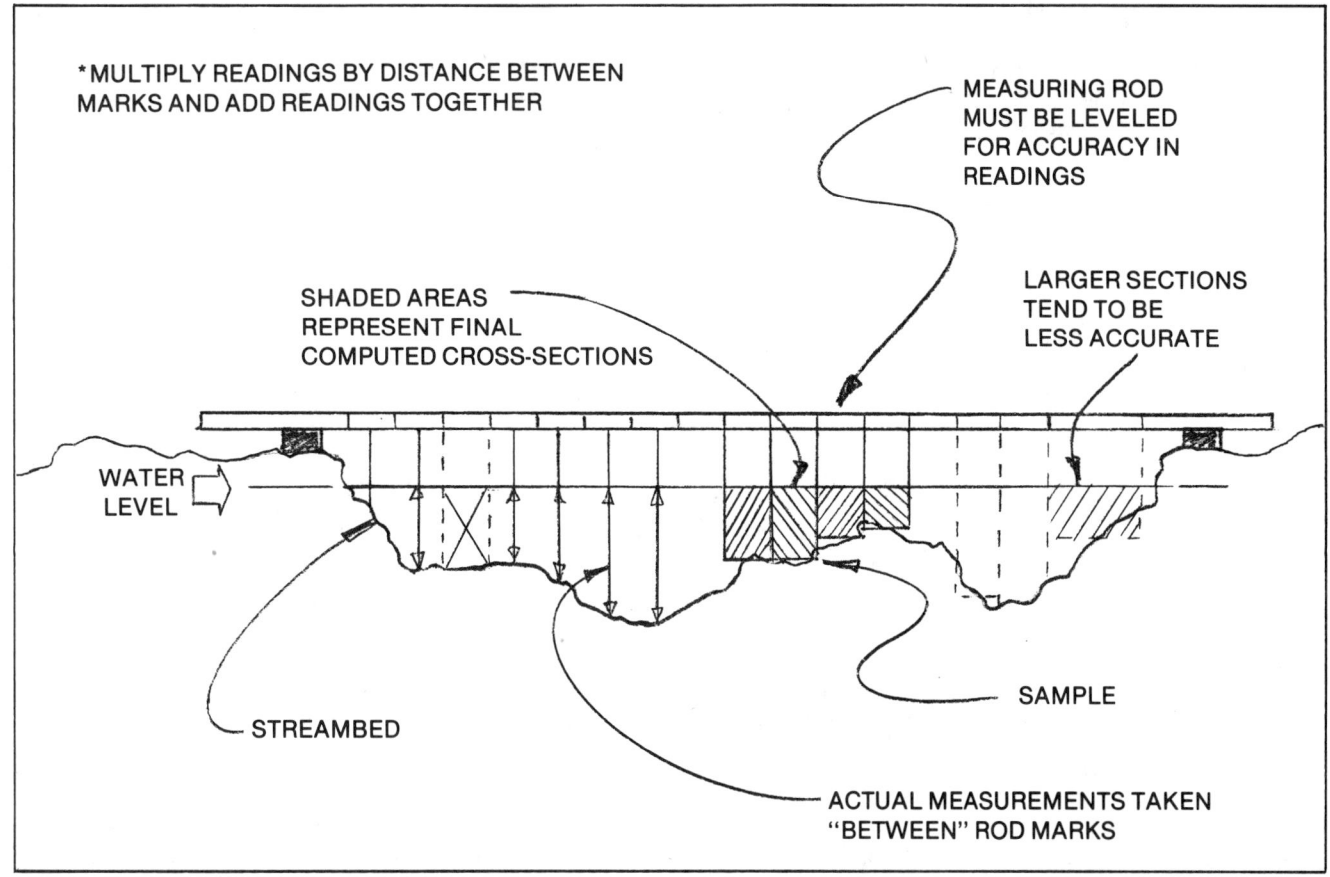

Fig. 6-9. Finding the cross-sectional area of a stream

Size it for the depth and velocity of the water flow, making it larger if it gets pushed in the direction of downstream; if it's hopeless, use a ruler edge-on to the flow and level to indicate that you're measuring straight down.

Whatever you use and however you do it, you're trying to find the depth of water at each point between the equally spaced markings on your rod. If you imagined that an invisible line extends down into the water from each of your markings, we'd have a rectangle with the top of the water as one side, the bottom another, and the invisible lines as sides. By noting the depth *between* the markings we pretend that's the length of the sides of our hypothetical rectangle. Multiply it times the distance between each of your markings on the rod and you have an area. You can't multiply inches by feet, so keep all of it in inches (and we'll convert to feet later) or in feet (or portions thereof). This is not the true cross-sectional area; it's the average cross-sectional area. Add all of the areas together and you've got a facsimile that's good enough first time around.

Your best bet is to pick a rock-free portion of the stream to take measurements; nobody needs to know how far down it is to a rock. If you can't find a good place in a few minutes, at least look for calm water and take your measurements. If you doubt your findings, take the average cross-sectional area of another point on the stream; if the water's velocity is the same, so must be the cross-sectional areas.

Speaking of the water's velocity, that's next. Get a bottle and put some small rocks or sand in the bottom until it sinks down a ways. If you've a companion along, he or she gets a bottle too—you're going to have a bottle race. It's a timed event; a stopwatch or secondhand on the wristwatch will do. It starts where you made your first measurement (with the rod) and ends at the second place you measured (Fig. 6-10). Or the starting gate is at a distance upstream of your measurement, and the finish is at the same distance downstream of the same measurement. Again, wherever or however, you should conduct your run near the measured point for the cross-sectional area. For the same quantity of water, a wide and shallow section has a different velocity than one that's deep and wide or narrow and shallow. Race where you've measured.

The sides and bottom of a stream slow the moving water, and so, with no friction to fight, the topmost water in the center of the stream has the greatest velocity. Therefore, measuring the time it takes for something floating directly on the surface to go a

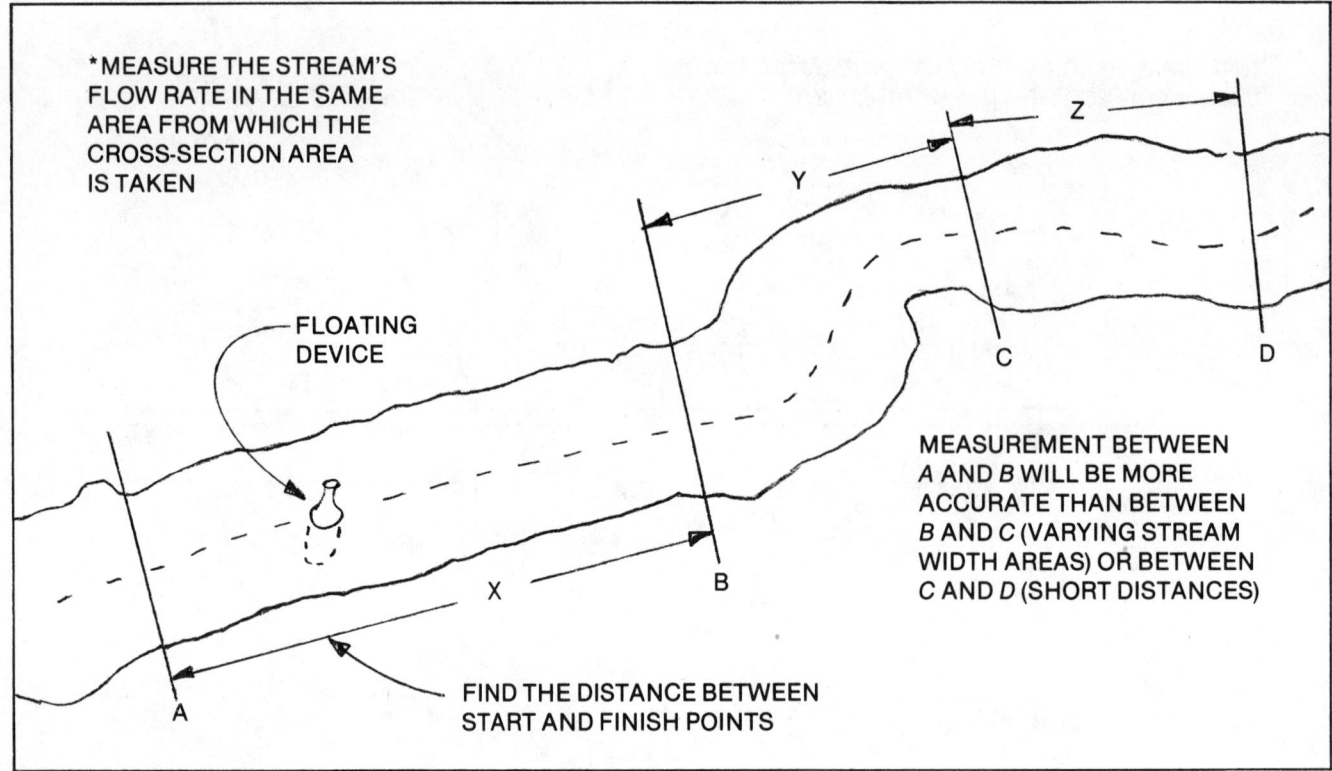

Fig. 6-10. Measuring the speed of water in a stream

certain distance is not going to represent the stream's average velocity accurately. Hence the weighted bottle. Irrespective of the stream's depth or width, something which is partly submerged will better portray the average speed of the water. If the stream is much deeper than the bottle's reach, get a bigger bottle. If the stream is violent and has big rocks, don't use a glass bottle. I don't really care if it breaks (and screws up your measurement), but whoever finds it with their bare feet will.

Speed is always described in reference to a distance (400 ft/sec or 18 mph), so we need a distance as well as a rate. Mark it off. The longer the better, *if* the depth and width of the stream stay constant. If not, find a shorter distance to match the width and depth of the cross-sectional area you picked. If it's real short—only a 5-second bottle run at most—you might pick two cross-sectional areas to check and, after running races in each, compare the two net readings. If significantly different, try again; if not, reduce them to the needed units.

Let's take an example. You've got a total (average) cross-sectional area of stream equal to 320 in^3 divided by 144 in^3 (the quantity of in^3 in a 1 ft^3) and we get 2.22 ft^3. After four consecutive races, we still get 2.6 seconds. So, 2.22 ft^3 in 2.6 seconds is identical with .85 ft^3 per second. To check this, you measure a cross-sectional area in a slower-moving portion of the stream and arrive at 680 in^3, which, divided by 144 in^3 per ft^3, equals 4.72 ft^3. And, at 5.4 seconds on the stopwatch, that's .87 ft^3 per second, which is close enough to .85 to call it a day. Better to be slightly conservative rather than optimistic (in some things), so we'll use the lower figure later on.

Keep at it until figures start getting close to each other. If the stream is borderline for useful power extraction, remember: you're going to commit yourself on this information. It's always a shame not to use power when it's there for the taking, but even more of a shame to install something as if the power were there, when it isn't. Unfortunately, because of the inherent limitations of this method, it's not nearly as accurate as you might want or need it to be. If that's the case, check out the next one.

Weir. Provided that you can get all of the stream's water to go through a specially built device (the weir), there's another way to compute, with astounding accuracy, the rate of flow (see Fig. 6-11). This method recognizes fluid flow and basic hydraulic principles, and with a simple chart (Fig. 6-12), you can magically transform your readings into flow rate.

Well, the measuring is easy. But you can't say the same for building the doggone thing. To build a weir, you first select a site with a smooth bottom and straight run into the weir location. There are the two sides and the dam bulkhead over which the water will flow. This notch into the dam must be at least three times as wide (L) as it is deep (H) for accuracy. You

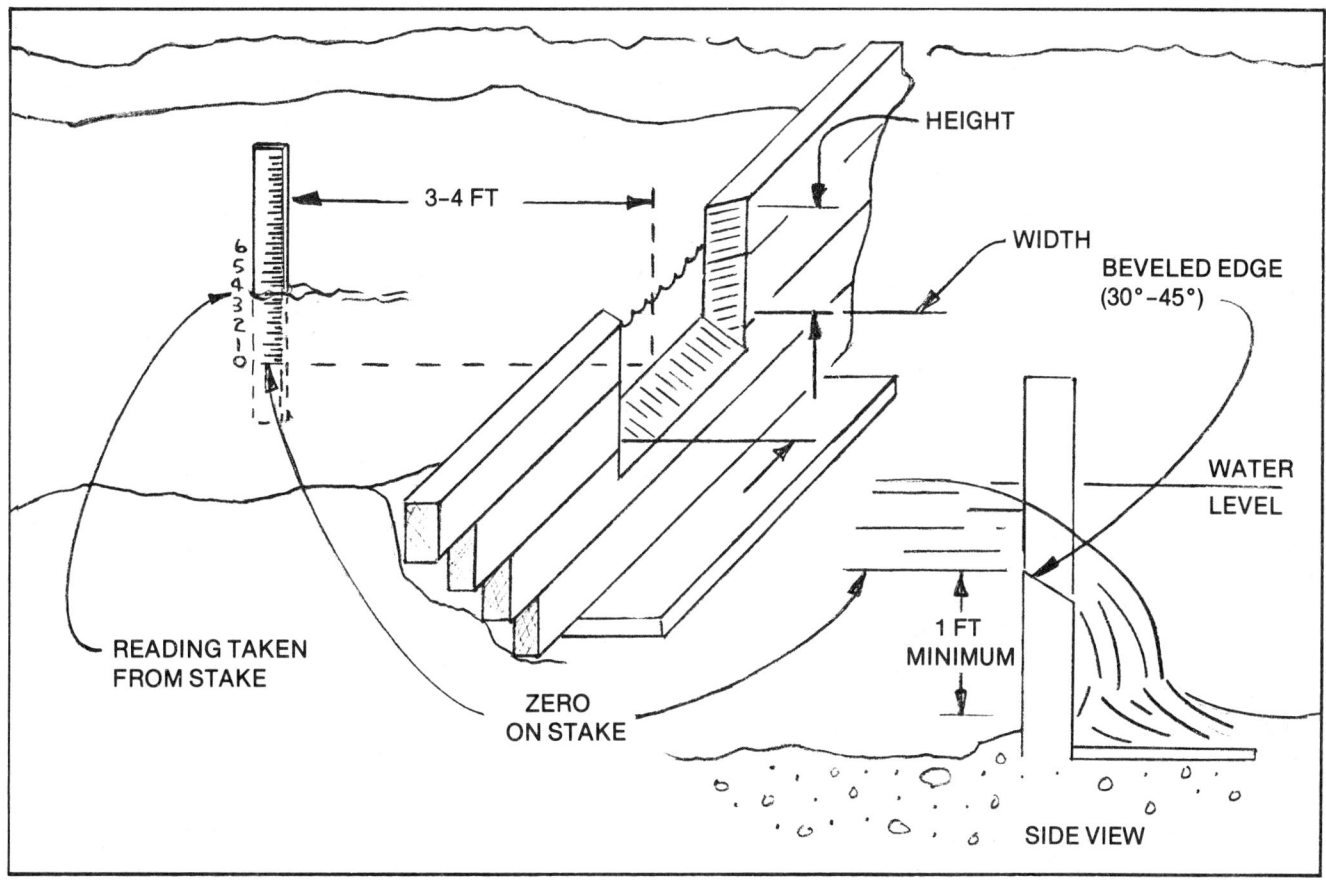

Fig. 6-11. The Weir method of finding flow rate

Depth (inches)	Flow rate ft³/sec/inch of Weir width
1	.007
2	.018
3	.035
4	.053
5	.075
6	.098
7	.123
8	.152
9	.180
10	.212
11	.243
12	.278
13	.313
14	.352
15	.388
16	.428
17	.468
18	.510

Fig. 6-12. Weir measurement table

don't want the water, in the most flooded condition of the stream, to be able to pass over the sides. *If you want to measure the maximum flow rate, that is.*

Building the weir is the nasty part of the job. Building it in sections, get it set so that no water can flow around or through the dam. The edges of the notch—sides and bottom—must be beveled with the knife edge on the upstream part. This is critical to accurate readings. The water flowing over the notch must not be interfered with by the tailwater below the weir; this is assured if there's a fall of one foot or more. The crest of the notch must be perfectly level.

We don't take any measurements at the weir itself. These occur upstream at a distance no less than four times H (notch depth); 3 or 4 feet is adequate. Drive in a stake (vertically) that's approximately centered on the weir but definitely not in turbulated water. If we think of the weir as a true dam, we want the stake in the backed-up portion of water rather than in the stream which feeds the dammed lake.

Hope you built everything with the least flow in the stream! Because now comes another critical part: the stake must be marked level with the crest. You see, when the water is flowing over the crest, the depth of the water in the reservoir behind the dam is not the same as the level of water directly above the crest. That's the reason we do the measuring upstream. So we need a reference, which is the sharp edge of the crest. Once the dam and weir are in place, you use a level to mark the upstream stake. If you were to fill the dam until water is just at the level of the crest, but not going over it, the measuring stake's mark should be at the waterline. This must be accurate—the precision of your rate of flow determination is directly proportional to the time you take to get it right.

Now, drive a nail partway into the stake at the crest mark so you can rest your ruler on it when taking measurements. Don't count on eyeballing it when even a few inches of water covers it. Or, better yet, tack a ruler onto the stake with its lowermost edge on the mark and you can leave it there for fast readings. Either way, put the ruler edge-on to the flow to minimize disturbing the water's surface, or your readings will be off. If you're leaving it there, you might even file or sand the edges, downstream and upstream, to a knife edge to cancel its effect on the water.

Painstaking as the installation is, it's worth the effort for the serious potential waterpower user. Any time of year, you can quickly take the readings off the measurement stake and, using a table, figure the rate of flow. Write it down—date, reading, and calculated flow. Other notes may be worthwhile—e.g., turbidity of the water or its cloudiness (grit and sediment). Note a recent rainfall; later, this will help you learn how your creek responds to deluges and how quickly. If you get a real groundpounder of a rain, you may want to take special precautions to ensure the safety of your rig, if applicable. Too much data allows you to pick and choose what's relevant, whereas too little can cause you lost time if, at a later date, you want to verify or modify the installation.

Before we move on, let's consider an example. You've taken a reading and it comes out at 5 inches. Looking at the chart under the depth column, 5 inches equals .075 ft^3 per second. That's *per inch of the weir's width (L)*. Your weir is 3½ feet, or 42 inches, wide. Multiplying 42 by .075 ft^3, you get 3.15 ft^3 per second. Easy, isn't it? And, if you did everything correctly, very accurate.

To recap, you find the depth of the water at the measuring stake, find that figure in the reading column, move across the chart to find its equivalent (in ft^3 per second per inch of weir), and multiply this figure by the width, in inches, of the weir. Keeping all the terms straight, inches cancel out and your answer is in ft^3 per second.

Now what? Hang in there; there's one more formula to go. Plug in the answer you've just gotten and it will tell you, depending on which of two formulas is more convenient for you, how much horsepower this much water will produce or how many watts of electrical power can be made. (See the section "Power from Water" in Chapter Four).

SECTION H: MEASURING THE HEAD OF WATER

This is the poor man's method, done without a transit or a crew of surveyors. Study the drawing carefully (Fig. 6-13) so that you get all the steps right; it's easy

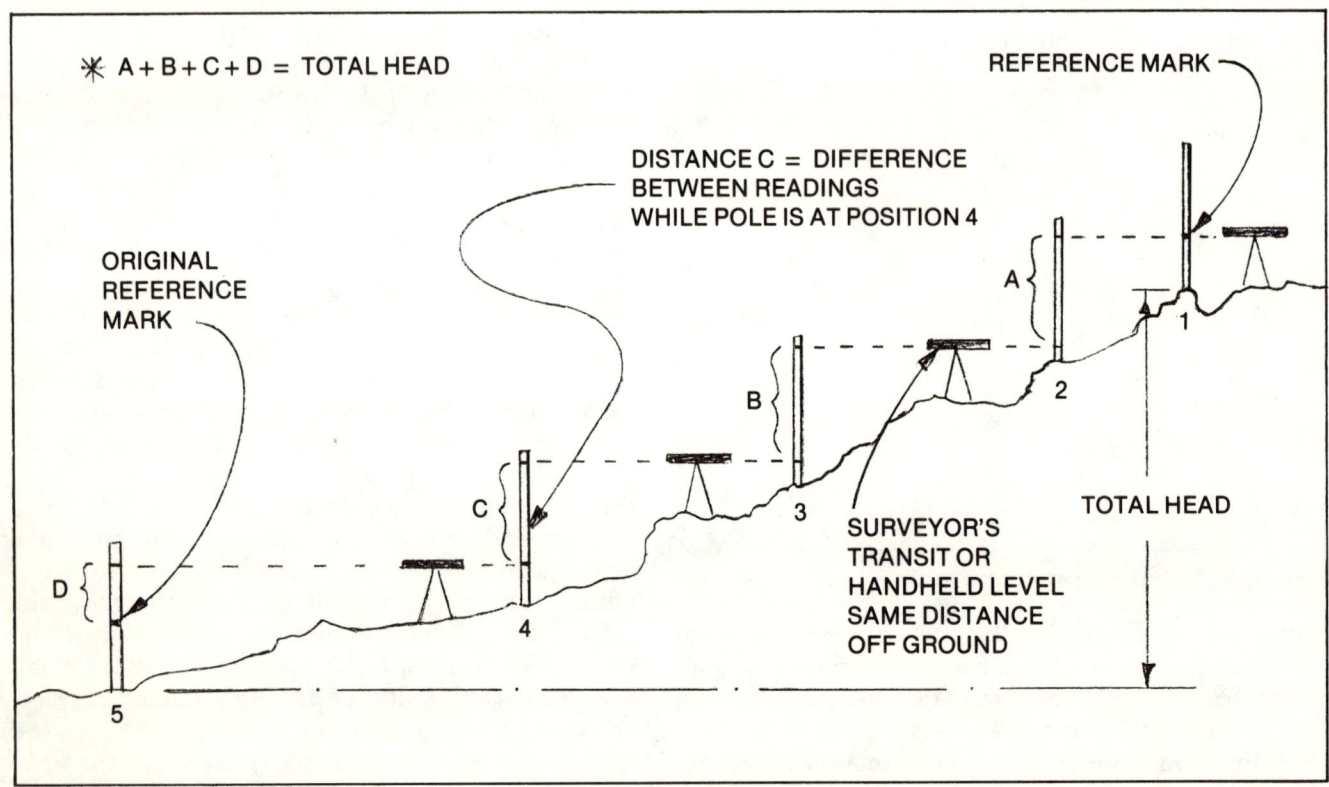

Fig. 6-13. Measuring a head of water

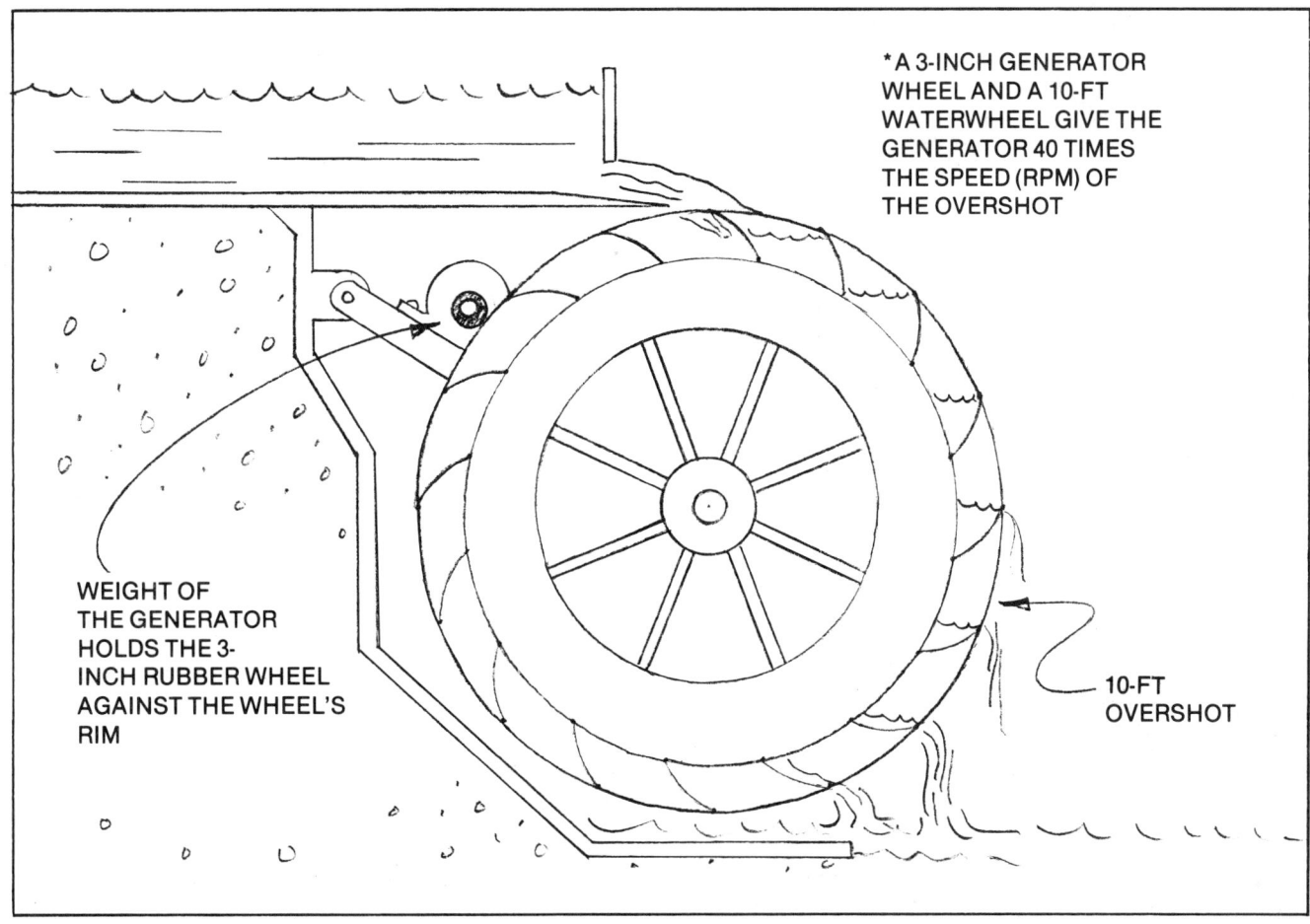

Fig. 6-14. A rim-drive hydroelectric unit

to get confused and add the wrong figures to the right ones.

SECTION I: AN ALTERNATE GEARING TECHNIQUE FOR WATERWHEELS

When the American Wind Turbine was introduced to the wind energy field a number of years ago, it brought a novel technique to the fore as well: rim drive. That is, while the center shaft is turning slowly and requires extensive gearing to achieve the necessary high rpm demanded by generators, the rim of the same windmachine is moving quite rapidly, because of its distance from the hub. Nowhere have I seen the same technique applied to waterwheels, which are inherently slow and plagued by the same problem if electrical generation is needed. Having never done this myself, I can't vouch that it will work, but here's the idea (Fig. 6-14). This can be a rim-pressure drive, where the generator has a rubber or steel "pulley" that is pressed against and turned by the waterwheel's rim, or an actual belt may be involved.

the generator and it rests against the outside of a

For example, we put a 3-inch-diameter wheel on 10-foot overshot (or Poncelet) wheel. The circumference of the waterwheel is 3.14 multiplied by 10 ft, or 31.4 ft. The circumference of the generator wheel is 3.14 multiplied by 0.25 ft (3 inches is 0.25 ft), or .785 feet. Divide the waterwheel's circumference by the generator wheel's circumference and the ratio comes out to an even 40:1. This can also be found by simply dividing the 10-foot wheel diameter by the 0.25-foot wheel diameter, also giving us a 40:1 ratio. Therefore, the waterwheel nets us 400 rpm at the generator. Not much if we need a few thousand rpm at the generator, but it's a start.

There are variations on this theme, but don't be naive in applying it. To transfer a lot of power by this method may not be practical—the wheel will slip just as a V-belt will if it doesn't have enough surface area. For installations under a few horsepower, however, it has excellent possibilities and should be considered. An intermediate jackshaft between the generator and waterwheel can make the next gear step; its ratio should be well under 10:1, which isn't difficult to come by. I wish you success.

SECTION J: CONVERSION TABLE

LENGTH

1 kilometer (km) = 0.6214 miles = 3281 feet
1 meter (m) = 1.0936 yards = 3.281 feet = 39.37 inches
1 centimeter = 0.0321 feet = 0.3937 inches

1 mile (mi) = 1.609 kilometer = 1609 meters
1 yard (yd) = 91.44 centimeters = 0.9144 meters
1 foot (ft) = 30.48 centimeters = 0.3048 meters
1 inch (in) = 2.54 centimeters = 0.0254 meters

MASS AND WEIGHT

1 kilogram (kg) = 2.2046 pounds
1 gram (g or gm) = 0.035 ounce

1 ton (2000 pounds) = 0.907 metric tons = 907.2 kilograms
1 pound (lb) = 0.4536 kilograms
1 ounce (oz) = 28.349 grams

VOLUME

1 cubic meter (m³) = 35.31 cubic feet
1 cubic centimeter (cm³) = 0.061 cubic inch

1 cubic yard (yd³) = 0.765 cubic meters
1 cubic foot (ft3) = 0.0283 cubic meters
1 cubic inch (in³) = 16.387 cubic centimeters

CAPACITY

1 liter (l) = 0.2642 gallon = 0.908 dry quart = 1.057 liquid quart = 0.03532 cubic foot

1 cubic foot (ft³) = 7.481 gallons = 28.32 liters
1 gallon (gal) = 3.785 liters = 0.1337 cubic foot
1 quart (qt) = 0.946 liters
1 pint (pt) = 0.473 liters

VELOCITY

1 kilometer per hour (km/hr) = 0.6214 miles per hour
1 meter per second (m/sec) = 3.281 feet per second = 2.237 miles per hour

1 mile per hour (mph) = 1.467 foot per second = 0.4470 meters per second
1 foot per second (ft/sec) = 0.3048 meters per second

TEMPERATURE

Centigrade (C) = $\frac{5}{9}$(°F − 32) = .555 (°F − 32)

Fahrenheit (F) = ($\frac{9}{5}$°C) + 32 = (1.8°C) + 32

MISCELLANEOUS

1 atmosphere of pressure (atm) = 14.7 pounds per square inch = 1.033 kilograms per square centimeter
1 horsepower = .7457 kilowatts = 550 foot-pounds per second
1 foot-pound (ft-lb) = 0.1383 kilogram/meter
1 British thermal unit (Btu) = 252 calories = 778.2 foot-pounds
1 kilocalorie (kc) = 3.968 Btu's
1 kilowatt hour (kwh) = 1.341 horsepower hour = 859.9 kilocalories = 3,412 Btu's

REALLY MISCELLANEOUS

40 poles = 1 furlong
1 hand = 4 inches
1 vara = 33.3 inches
2 stone = 28 pounds
1 scruple = 20 grams

Fig. 6–15. In a cycle, what looks like the end is really the beginning.

Sources and References

I never read the references sections in other people's books. Probably because it bores me. I don't really care to know about every book ever written on a subject. What I would like to know is: which are the *best*? Sure, such judgments are subjective, biased, and may step on some toes. But you can bet an author tells his friends! Nevertheless, if I shell out some hard-earned money for a book, I'd like some candid advice. So I'll practice what I preach. What you're getting in the next few pages is the *cream* of my library—in my opinion. I may not have all the cream —I'd be naive to think I'd seen 'em all—but I'm sharing what I do have with you. I judge a book by how much it gets used; believe me, these look as though an auto mechanic had been searching through them in the middle of a workday.

Many of the people doing good work in the alternative energy field are in small groups (a few people) or are borderline poverty-stricken. They don't have advertisement write-offs, postal budgets, or time to tell you everything they know about solar energy. Temper your enthusiasm with a little respect, forethought, and sensitivity. You want a brochure? Drop in a quarter or a few stamps with that request. Want some information? Make it worthwhile for the person who gets your letter to steer you on to a good source or reference, or to give you some experience-earned advice. Drop in a dollar or two; how much depends on how fast you want the reply and how good you want its contents to be. Don't forget the SASE (self-addressed, stamped envelope).

GENERAL

Alternative Sources of Energy Magazine. Quarterly. Back issues $2 each, subscription info from A.S.E., Rt. 2, Box 90A, Milaca, MN 56353. Clear do-it-yourself articles on using solar, wind, water, methane—projects and plans.

Appropriate Technology Sourcebook. By Ken Darrow and Rick Pam. Volunteers in Asia Publications, 1976. $4 from Box 4543, Stanford, CA 94305. Good energy bibliography.

Cloudburst—A Handbook of Rural Skills and Technology, and *Cloudburst 2*. Edited by Vic Marks. Cloudburst Press, 1974, 1976. $5.20 each from Cloudburst Press, Mayne Island, B.C., Canada VON 2JO. Do-it-yourself alternative energy and homestead building projects.

Earthbooks Lending Library, Sweet, ID 83670. Mail-order lending library of excellent books on energy, building, gardening, self-sufficiency. 50¢ for catalog.

Earthmind, 4844 Hirsch Road, Mariposa, CA 95338. Alternative energy books, slide shows, and services. Send 50¢ and long SASE for publications list.

Electric Vehicles—Design and Build Your Own. By Michael Hackleman. Peace Press, 1977. $9 from Earthmind (above).

Energybook #1, Energybook #2—Natural Sources and Backyard Applications. Edited by John Prenis. Running Press, 1975, 1977. $5 each in bookstores or from Running Press, 38 S. 19th St. Philadelphia, PA 19103 (add 25¢ postage).

Energy Primer—Solar Water, Wind, and Biofuels. Portola Institute, 1974. $4.50 in bookstores.

LeJay Manual. $3.50 from LeJay Mfg. Co., Belle Plaine, MN 56011. Lots of old-timey energy projects, complete plans, fun!

Living with Energy. By Ronald Alves and Charles Milligan. Penguin Books, 1978. $5.95 in bookstores.

Other Homes and Garbage—Designs for Self-Sufficient Living. By Leckie, Masters, Whitehouse, and Young. Sierra Club Books, 1975. $9.95 in bookstores or from Sierra Club Books, 530 Bush St., San Francisco, CA 94108.

Pedal Power—In Work, Leisure, and Transportation. Edited by James C. McCullagh. Rodale Press, 1977. Order from Rodale Books, Emmaus, PA 18049.

James L. Ruhle & Associates, P.O. Box 4301, Fullerton, CA 92631. Excellent color slide collections for sale on energy topics. Catalog.

Village Technology Handbook. By VITA Volunteers, 1970. Contact for availability: VITA, College Campus, Schenectady, NY 12308. Very simple, effective energy and water projects.

ENERGY CROSSROADS

Critical Mass Journal. Monthly. $7.50 from P.O. Box 1538, Washington, D.C. 20013. Keep informed on the anti-nuke movement—what's happening and where.

Energy. By John Holdren and Phillip Herrera. Sierra Club Battlebook. The best exposé on nuclear energy and other energy sources in use today.

SOLAR

Alternative Sources of Energy Magazine, No. 34, Special Edition: "Solar Hot Water."

A to Z Solar Products, 200 East 26th St. Minneapolis, MN 55404. Catalog.

A Bibliography for the Solar Home Builder. By Donald Aitken. California Office of Appropriate Technology, 1979. Order from OAT, 1530 Tenth St., Sacramento, CA 95814.

California Solar Information Packet. 1977. Contact for availability: California Energy Resources Conservation and Development Commission, 1111 Howe Ave., Sacramento, CA 95825.

Direct Use of the Sun's Energy. By Farrington Daniels. Ballantine Books, 1964. $2.50 from Earthmind. *The classic work.*

Energy in Building Design. By A. Davis and R. Schubert. 1974. Contact for availability: Passive Energy Systems, P.O. Box 499, Blacksburg, VA 24060.

The First Passive Solar Catalog. By David Bainbridge. Passive Solar Institute. $5 from P.O. Box 722, Davis, CA 95616.

The Food and Heat Producing Solar Greenhouse. By Rick Fisher and Bill Yanda. John Muir Publications, 1976. $6 in bookstores or from Bookpeople, 2940 Seventh St., Berkeley, CA 94710.

Heat-transfer cements: Chemax Corp., 211 River Rd., New Castle, DE 19720. Write for catalog. (We used "Tracit.")

How to Keep Your House Warm in Winter, Cool in Summer. U.S. Dept. of HUD Cornerstone Library. $2.95 in bookstores.

Natural Solar Architecture—A Passive Primer. By David Wright. Van Nostrand Reinhold, 1978. $7.95 in bookstores or from Litton Educational Publishing, 136 W. 50th St., New York, N.Y. 10020.

New Sources of Energy, Volume 5: Solar. Proceedings of U.N. Rome Conference 1961, United Nations Publications. $16 from United Nations, Sales Section, New York.

Practical Solar Heating. By Kevin McCartney. Prism Press, 1978. $3.95 in bookstores or from Prism Press, Stable Court, Chalmington, Dorchester, Dorset DT2 0HB, England. Excellent!

Practical Sun Power. By William H. Rankins and David A. Wilson. Lorien House, 1974. $4 from P.O. Box 1112, Black Mountain, NC 28711. *Very* workable projects.

Resource Conservation for Residential Buildings. Project Director, Gary Starr. Contact for availability: Gary Starr, 1675 Escalante, Burlingame, CA 94010.

Sizing of solar energy systems: see *Natural Solar Architecture, The Solar Home Book, Other Homes and Garbage.*

The Solar Cookery Book. By Beth and Dan Halacy. Peace Peace, 1978. $6.95 in bookstores or from 3828 Willat Ave., Culver City, CA 90230. Detailed plans for Halacy oven and other solar cooking devices, plus twenty years of solar cooking recipes. Excellent drawings, photos, and designs.

Solar Energy Handbook. By Henry Landa, 1974. Contact for availability: F.I.C.O.A., 2901 S. Wentworth Ave., Milwaukee, WI 53207.

National Solar Heating and Cooling Information Center, P.O. Box 1607, Rockville, MD 20850.

Solar Energy and Shelter Design. By Bruce Anderson. 1973. Contact for availability: TEA, Box 47, Harrisville, NH 03450.

The Solar Home Book. By Bruce Anderson. Cheshire Books, 1976. $7.50 in bookstores or from Cheshire Books, Church Hill, Harrisville, NH 03450. Excellent!

Sunspots. By Steve Baer. Zomeworks, 1975. $4 from P.O. Box 712, Albuquerque, NM 87103.

The Survival Greenhouse—An Ecosystem Approach to Home Food Production. By James B. DeKorne. Peace Press, 1975. $8 from Earthmind. Excellent!

SOLAR REFRIGERATION

Alternative Sources of Energy Magazine, No. 14 (May 1974), "Natural Water Cooling and Freezing."

Domestic Heat Pumps. By John Sumner. Prism Press, 1976. Approximately $6 in bookstores or £2.95 from Prism Press, Stable Court, Chalmington, Dorchester, Dorset, DT2 OHB, England.

"An Experimental Intermittent Absorption Refrigerator." By Robert K. Swartman. 1968. Contact author at Faculty of Engineering Science, University of Western Ontario, London, Ontario, Canada.

"How Solar Heat Can Cool Your Home," *Popular Science*, Sept. 1975, p. 68.

"Intermittent Absorption Refrigeration." By Don Marier. *Alternative Sources of Energy Magazine*, No. 20 (March 1976), p. 7. The icy-ball type.

Servel Gas Refrigerator Manual. Operation, maintenance, parts, conversion valve sizes, repairs. Out of print, but Earthmind can xerox its library copy for you. Write for price.

"Solar Powered Refrigerator." By R. Swartman. *Mechanical Engineering Magazine*, June 1971, p. 22.

WIND

Burstein-Applebee Electronics, 3199 Mercier St., Kansas City, MO 64111. Parts catalog, $2.

Delco Engine Manual. $10 from Earthmind. No longer available from Delco. Complete parts, installation, operation, and service info for Delco standby generators 1932–39.

Delco Sentry Manual. $3 from Earthmind. No longer available from Delco. Technical manual for Sentry control box, for automatic standby generator startup, operation, and shutdown.

Dwyer Instruments, Michigan City, IN 46360. Windspeed indicators.

The Homebuilt, Wind-Generated Electricity Handbook. By Michael Hackleman. Peace Press, 1975. $9 from Earthmind. Tower-raising, wind-electric machine restoration and installation, design notes, standby generators.

Jacobs Manual. $3 from Earthmind. A reprint of the original installation and operator's manual accompanying a Jacobs wind-electric system.

New Developments in Homebuilt Windmill Technology. By Kevin Varner. 1979. $10 from the author, RR 2, Ogilvie, MN 56358.

Octahedron-module tower plans. Contact Earthmind for more information.

Fig. 7-1. Here's Vern with some music to liven up an otherwise dull section.

R.A. Simerl Instrument Div., 238 West St., Annapolis, MD 21401. Windspeed indicators.

Wincharger Manual. $3 from Earthmind. Same as *Jacobs Manual*, only for Winchargers.

Wind and Windspinners—A "Nuts-'n-Bolts" Approach to Wind-Electric Systems. By Michael Hackleman. Peace Press, 1974. $9 from Earthmind. Basics of wind, electricity, batteries, and step-by-step building of an S-rotor.

Wind in California. Dept. of Water Resources, 1978. $3 from State of California, Dept. of Water Resources, P.O. Box 388, Sacramento, CA 95802. Windspeed data from all weather stations in California.

Windlight Workshop, Rt. 2, Box 271, Santa Fe, NM 87501. Restoration and installation of "antique" wind-electric systems and wind water pumpers. Tower-raising, spare parts, electronic controls, and load diverters. $1 for info.; ask for "Windy."

Wind Power Digest. Quarterly. 109 East Lexington, Elkhart, IN 46514. $6 per year. Back issues from Earthmind. Best windpower magazine ever.

WOOD

Alternative Sources of Energy Magazine, No. 35, Special Edition: "Wood Heat."

Handmade Hot Water Systems. By Art Sussman and Richard Frazier. Garcia River Press, 1979. $4.95 from P.O. Box 527, Point Arena, CA 95468. Also sell "Blazing Showers" wood-fueled water-heating systems.

"Revitalizing Wood Ranges" by Kurt Boyer. *Cloudburst II*, p. 38.

Wood-Burning Quarterly and Home Energy Digest. $7.95 subscription from 8009 34th Ave. S., Minneapolis, MN 55420. Wood-heating your home, all kinds of stoves, woodlot management, tools and tips, small energy projects—a good magazine!

Wood-burning water heaters: Appropriate Technology Importers, P.O. Box 5, El Rito, NM 87530.

WATER

The Banki Water Turbine. C.A. Mockmore and Fred Merryfield, Bulletin Series #25, February 1949. $1.00 from Engineering Experiment Station, Oregon State College, Corvallis, Oregon 97331. A must-have, complete set of plans and specifications for an owner-built hydroelectric unit.

Building a hydraulic ram: three designs in *Cloudburst II.*

Cloudburst 1. Waterpower info: general design notes, Banki turbine, overshot and undershot wheels.

Energy Primer. Waterpower info: Pelton wheel construction, Banki turbine, overshot wheel design.

Harnessing Water Power for Home Energy. By Dermot McGuigan. Garden Way Books, 1978. $4.95 from Garden Way, Charlotte, VT 05445.

Pharaoh's Pump. By Edward Kunkel. 1962. $3.50 from the author, 295 W. Market St., Warren, OH 44481. Unique approach to pumping water using hydraulic-ram principles.

Water Watts hydroelectric generator: info from Electrical Independence Co., P.O. Box 1278, Willits, CA 95490.

"Reciprocating Wire Power Transmission for Small Waterwheels," *Village Technology Handbook*, p. 117.

METHANE

The Compleat Biogas Handbook. By D. House. 1978. $8 from At Home Everywhere, c/o VAHID, Rt. 2, Box 259, Aurora, OR 97002.

Practical Building of Methane Power Plants. By L. John Fry. 1974. $12 from Mother's Bookshelf, P.O. Box 70, Hendersonville, NC 28739. Best methane book we've seen so far.

Stop the Five Gallon Flush—A Survey of Alternative Waste Disposal Systems. 1973. $2 (Canadian) from Minimum Cost Housing Group, School of Architecture, McGill University, P.O. Box 6070, Montreal H3C 3G1, Canada.

Index

Numbers in italics refer to illustrations

AAE (average annual energy), 63
AAW (average annual windspeed), 63
Absorber collector
    advantages/disadvantages of, 16–19
    air-medium types of, 11–13
    characteristics of, 9–13, 36–37
    expansion coefficient in, 23
    water-medium types of, 9–11
    (*see also* Collector)
AC (alternating current), 75–76
Active systems
    of solar utilization, 8
    of space heating, 31
Adobe Trombe wall, *30*
Aerobic decomposition, 116
Aeroelectric machine, 60 (*see also* Wind energy systems)
Aeroturbine
    area of, 62
    classes and types of, 65–74
    control of, 78–83, *83*
    effect of turbulence on, 86–87
    efficiency, 62, 128, *129*
    ratings for, 74–75
    and wind energy, 60–61
    (*see also* Wind energy systems)
ah (ampere-hour) rating, 76
Air density, 62
Airfoils, 67
Air-medium collectors, 11–13, *11*, *12* (*see also* Absorber collector)
Air-medium heat transfer
    advantages/disadvantages of, 13
    collectors for, 11–13
    prerequisites for, 23, 27
Air-medium storage, 27
Air-medium thermosiphoning, 23
Aluminized Mylar (*see* Mylar)
Ammonia as refrigerant, 56 (*see also* Cooling techniques)
Anaerobic digestion, 120
Angle of declination, 32–34, 127, *127*, *128*
Antifreeze
    in collectors, 23–24
    in icy-ball refrigerator, 56
Aqueduct (*see* Sluice)
Asbestos in wood-burning stoves, 101
Auxiliary power (*see* Standby systems)

Backup systems (*see* Standby systems)
Baer, Steve, 28, 29, 50
Banki turbine, 107–8, *108* (*see also* Michell turbine)
Batch load digester, 120, *121*
Battery
    arrangement, 77–78, *77*
    capacity, 76
    characteristics, 76–78
    rating for wind energy systems, 76–78
    use in standby systems, 86
    use in water energy systems, 112–13
    use in wind energy systems, 75, 76–78, 83, 112
    voltage, 77–78
Beadwall, 50, *50* (*see also* Insulation)
Biogas
    composition of, 116–17
    reservoir for, 121, *121*
    (*see also* Methane)
Blades (*see* Propellers)
Blowers, 23, 27
Bonding techniques (*see* Thermal bonding)
Braking in aeroturbine, 81
Breast wheel, 106
Building codes
    for methane energy, 122–23
    for solar energy, 24, 43–45, 58
    for water energy, 115–16
    for wood energy, 101

Cement, heat transfer, 10 (*see also* Thermal bonding)
Chain saw, 101–2
Check valve, 20, 21, 91, 105
C/N (carbon to nitrogen) ratio, 119
Coating for absorbers, 36
Codes (*see* Building codes)
Collector
    absorber, 9–13, 16–19, 36–37
    characteristics of, 36–43
    concentrator, 13–16, 16–19
    condensation in, 37
    construction of, 36–37, 51–52
    dissimilar metals in, 41
    ducting in, 25
    efficiency factors of, 16–19
    functions in a solar utilizing system, 5–6
    losses at the, 35–36

mounting of, 44–45, *44*
orientation of, 16, 18, 19, 33–34, 127
siting of, 58
sizing of, 34
types of, 9–19
variable conditions in construction of, 5–6
(*see also* Absorber collector; Concentrator collector)
Combustion air, 99–101
Concentrator collector
advantages/disadvantages of, 16–19
types of, 13–16
(*see also* Collector)
Concrete in heat storage, 30 (*see also* Trombe wall)
Condensation in collectors, 37
Conduction, 6–8, *7*, 42 (*see also* Insulation)
Continuous feed digester, 119–20, *120*
Control
in solar energy systems, 58–59
in water energy systems, 112
in wind energy systems, 78–83, *79*, *82–83*
Convection, 6–8, *7*, 52–53
Cooking
energy conservation in, 125
solar, 45–48, 125
with wood heat, 99–100
Cookstoves, wood-burning, 99–100, *100*
Cooling techniques
convective, 52–53
evaporative, 31, 53–54
exposure, 52
heat pump, 55–56
night-sky radiation, 54–55
refrigeration, 52
shading, 53
Cord of wood, 97
Corrugated sheet, 11
Cost
of glazings, 41–42
of solar usage, 59
of towers, 88
of windplants, 92–94
Creosote in wood stoves, 101
Crossflow turbine (*see* Michell turbine)
Cube law in wind energy, 63

Dams, 111–12
DC-only systems
in water energy, 113
in wind energy, 77, 83–84
Declination angle, 32–34, 127, *127*, *128*
Deep-cycle batteries, 76–77
Digester, 118–23 (*see also* Methane)
Dish, parabolic, 13, 15, *15*
Double-action collector, 12, *12*
Drip collector, 10–11, *10*
D-rotor (Darrieus rotor), 66, 70, 72–73, *72*, 129
Drumwall, 28–30, *28*
Dual-fuel stoves, 100
Ducting in solar collector, 25
Dynamo, 67, 70

Earth as insulation, 43
Efficiency
of aeroturbines, 73–74, 128, 129
of fireplaces, 98–99
of inverters, 84
of propeller-type windplants, 68–69
of waterwheels and water turbines, 109
in wind energy conversion, 64–65
Eggbeater (*see* D-rotor)
Electricity
conversion to horsepower, 110–11
storage of, 75–78, 112–13
from water energy, 105, 110
from wind energy, 61–65, 79–80, 83–86, 129
Energy conservation
in cooking, 125
in refrigeration, 124
in utility usage, 4
in wind systems, 63–65, 85
Energy winds, 61
Eutectic salts, 26–27
Evaporative cooling, 31, 53–54 (*see also* Cooling techniques)
Expansion coefficient
in absorbers, 23
in glazings, 41, 42
Exposure, 52 (*see also* Cooling techniques)

Farm windmill (*see* Water pumping)
Feathering (*see* Spoiling)
Fiberglass, 36, 42–43 (*see also* Insulation)
Filters, biogas, 116
Firebox rebuilding in stoves, 100, *100*
Fireplaces, 98–99
Flat-plate Collector (*see* Absorber collector)
Flat reflector, 13, 15, *17* (*see also* Concentrator collector)
Flow rate calculation, 103, 130–34, *131*, *132*
Focal distance, 16, 18–19, *19*
Focal tube, 16, 18–19
Food dryer, solar, 48–49
Forestry permits for woodcutting, 96
Freeze protection in solar usage, 23–25
Fresnel lens, 13, 14–15, *14* (*see also* Parabola)
Furling of windplants, 83

Galvanic action, 41
Gassing in batteries, 78
Gearing for waterwheels, 135
Generators, standby, 125
Geothermal, 3
Glass
for fireplaces, 99
as glazing material, 37, 41
thickness in glazing, 42
Glass lens (*see* Parabola)
Glauber's salt, 26–27
Glazing
heat losses from, 36–37
materials and specifications for, 41–42
Glycol, 23–25
Governor, windplant, 79–82, *82*
Greenhouse effect, 41, 49–50
Greenhouses, 49–50, *49*
Guyed tower, 93, *93* (*see also* Towers)
Gyroscopic vibration, 68, *68*

Halacy oven, 46–48, *46*, *47*, 125
Hardwood, 98
HAW (horizontal axis windplant)
orientation of, 65–66
sliprings for, 66–67, *66*, *67*
types of, 67–70
Hay, Harold, 30

Head
    measurement of, 134–35, *134*
    in water energy systems, 103, 109, 114
Headrace in dams, 111
Heat
    capacity, 8–9
    conduction of, 6–8, *7*
    exchange, 23–25
    of fusion, 25
    lag, 35
    latent, 25
    materials for storage of, 25–27
    paths, 6–8
    pumps, 55–56
    radiation of, 6–8, *7*
    specific, 25–27
    thermal, *7*
    (*see also* Heat-transfer)
Heat-transfer
    bonding techniques for, 10
    function of medium for, 6
    medium of air, 11–13, 23
    medium of water, 10–11, 19–23
    in solar energy system, 5–6
Heating stoves, wood, 99
Heliostat, 16, *17*
Hindenburg effect, 78
Horsepower
    conversion to kilowatts, 110–11
    in water energy systems, 108–10
Hubbard squash, 40, *40*
Hybrid aeroturbine, 70, 73, *73*, 129
Hydraulic ram, 104–5, *104*, 114 (*see also* Water pumping)
Hydrogen sulfide, 116–17
Hydrometer, 78
Hydrothermal, 3

Icy-ball refrigerator, 56, *57*
Impact resistance of glazings, 42
Impulse turbines in water energy, 107–8
Insulation
    beadwall, 50, *50*
    of the collector, 36–37
    vs. conduction, 6, 42
    fiberglass, 36
    materials, comparison of, 42
    in refrigeration, 124–25
    in solar energy systems, 42–43
Inverters, 89

Jacobs control box, *79*
Jacobs generator, *89*
Jacobs governor, *80*

kw (kilowatt)
    conversion to horsepower, 110–11
    value in water energy, 110
    value in wind energy, 63, 129
kwh (kilowatt-hour) computation of, 84–85

Legality of systems (*see* Building codes)
Litmus paper, 118–19
Loading in governing, 81–82
Lollyshaft (*see* Turntable)

Manual shutdown of windplants, 82–83
Mechanical energy from waterpower, 104

Medium transfer, types of, 19–25 (*see also* Thermosiphoning; Water pumping)
Methane
    and biogas, 116–17
    compression of, 117, *118*
    digester, 118–23
    energy yield from, 122
    and fertilizer, 120
    generators of, 120
    sources of, 3, 116
    storage of, 121
    uses for, 117
Metrics, use of, 1–2
Michell turbine, 107–8, *108*, 114
Mylar, aluminized
    as glazing material, 41
    in solar oven, 46, *47*
    as vapor barrier, 37

Night-sky radiation, 11, 31, 54–55 (*see also* Cooling techniques)

Octahedron module tower, 92, 93, *93*
Orientation
    of solar collectors, 16, 18, 19, 33–34, 127
    in solar cooking, 47, 125
    of windplants, 65–66
Overshot wheel, 106, 107, *107*, 114, 135, *135*

Parabola, 13–14
Parabolic units, 14–15, 45, *128*
Passive systems in solar utilization, 8, 27–31
Pelton wheel, 107, 108, *109*, 114
pH in digesters, 118–19
Pinson cycloturbine, 73
Piping
    in methane energy systems, 121–22
    in solar energy systems, 25
    in water energy systems, 112
Poncelet wheel, 107, 135, *135*
Pool collector, 11
Power ratings
    in water energy systems, 103–4
    in wind energy systems, 62, 74–75, 92
Pressure, water, 103–4
Prevalent winds, 61, 88
Primary systems of solar water-heating, 23, *23*, 51
Propeller-type aeroturbine, 67–69, *71*
Propellers
    efficiency of, 129
    gyroscopic vibration of, 68
    taper/twist in, 68–69
Pump jack, *90*, 91
Pumps, 21–23 (*see also* Water pumping)
Pump standard, 91

Radiation, 6–8, *7* (*see also* Night-sky radiation)
Ram, hydraulic, 104–5, *104*, 114 (*see also* Water pumping)
Reflectivity, 8, 34
Reflectors, 47, 125
Refrigerants, 55–56
Refrigeration
    application of, *124–25*
    devices used for, 56
    principles of, 55–56
    (*see also* Cooling techniques)
Refrigerator, night-sky, *54*

Rim drive of waterwheel, 135
Riser/header, 39–40, *39*
Rock
  for heat storage, 26–27
  specific heat of, 25
Roof loading, 43
Rooftop cooling, 54 (*see also* Cooling techniques)
R-value of insulation, 42, 50

Safety
  with batteries, 78
  with concentrators, 19
  in methane use, 117
  in solar usage, 58
  in woodcutting, 102
  in wood energy usage, 101
Salts for heat storage, 25, 26–27
Sanitation (*see* Building codes)
Seasonal variations
  of solar energy, 32–33
  of streams and rivers, 103
Secondary systems
  of water heating, 23–25
  for wood energy, 95
SGU (standby generator unit), 85–86, 94 (*see also* Standby systems)
Shading, 53
Side-facing
  for manual shutdown, 82–83
  windmachines, 81
Silting in dams, 111
Single-action collector, 12
Siting
  of solar collectors, 34–35, 43–44
  of solar storage, 58
  of towers, 86–88
  of water energy systems, 115
Sizing
  for application, 27
  of methane digester, 122
  of solar collector tray, 19
  of solar energy systems, 27, 34, 126
  of solar glazing, 42
  of wind energy systems, 65, 84, 91, 93
Skylid, 29
Skytherm, 30–31, *31*
Sliprings in windplant, 66, 67, *67*
S/L ratio, 119
Sluice in dams, 111
Softwoods, 98
Solar barbecue, 45–46 *46* (*see also* Solar cooking)
Solar checklist, 58–59
Solar collector (*see* Collector)
Solar cooking, 45–48, 125
Solar energy
  availability of, 3, 31–36
  characteristics of, 3
  collection of (*see* Collector)
  general uses for, 5
  specific uses for, 45–52
  storage of, 6
  (*see also* Solar utilization systems)
Solar food dryer, 48–49, *48*
Solar heat, 11, 126 (*see also* Heat)
Solar light, 7–8
Solar still, *48*
Solar tracking (*see also* Orientation)

Solar utilization systems
  active, 8
  building codes for, 58
  functions of, 5–6
  medium transfer in, 21–25
  passive, 8
  parts of, 5–6, *6*
  standby, 126
  tax credit for, 59
Solar water heating (*see* Water heating)
Space-heating systems
  active, 31
  air-medium, 13
  passive, 27–31
  window-collector, 12
  wood-burning, 99
Specific heat in storage materials, 25–27
Spoiling in windplants, 80
S-rotor (Savonius rotor), 66, 70–72, *71, 72, 89, 91,* 129
Stalling in windplants, 81
Standby systems
  for solar energy, 45, 126
  for wind energy, 85–86, 94, 125
Starter in digesters, 122
Storage
  for application temperatures, 27
  heat lag in, 35
  in methane systems, 120–21
  sizing of, 27
  in solar energy systems, 5–6, 23, 25–27, 126
  in water energy systems, 113–14
  in wind energy systems, 75–76
Storage tank
  in methane energy systems, 120–21
  in pumped water systems, 21
  as standby water source, 91
  in thermosiphon system, 20–21
  for water, 26
Stoves, 99–100
STP (Standard temperature and pressure), 55
Stuffing box (*see* Pump standard)
Sunshine law, 44
Synchronous inverter, 75
Swamp cooler, 53–54 (*see also* Cooling techniques)

Tax credit
  for solar energy, 59
  for wind energy, 94
Temperature
  control by sizing, 19
  differentials in collectors, 36
  in digesters, 119
  in solar energy systems, 27
  in thermosiphoning, 22–23
T&P (temperature and pressure) valve, 20, 37, 51
Thermal bonding
  cements, 10
  techniques, 10, 40–41
  in thermosiphoning, 23
Thermal heat, 7
Thermal mass, 36, 47
Thermal shock, 23, 41
Thermosiphoning
  air-medium, 23
  in masonry fireplaces, 100
  process and systems of, 19–25
  in tube-on-sheet collectors, 38

in water heating, 51–52
Towers, 86–89, *87*, 93, *93*
Tracking (*see* Orientation)
Transmittance, 7, 8, 41
Tray collectors, 13, 15, *16*, 18, *19* (*see also* Collector; Parabolic units)
Trombe wall, 5, 30, *30*
Tube-in-sheet collector, 10
Tube-on-sheet collector, 9, 10, *10*, 37–41
Tubed-sheet collector, 10, *10*
Turbines
    air (*see* Aeroturbine)
    water, 107–8, 114
Turbulence, air, 86–87
Turntable, windplant, 66–67
Twin-sheet collector, 9, *9*

Ultraviolet (UV) resistance of glazing material, 41
Undershot wheel, 106
Universal motor, 83
u/V ratio for propellers, 68

VAW (vertical axis windplant), 70–73
Venturi shroud, 66, 70, *71*
Voltage in battery systems, 77–78

Waste disposal in methane energy, 122–23
Water
    flow rates of, 103, 104–5, 130–32
    for generating electricity, 105
    as heat storage material, 25–26
    as heat-transfer medium, 9–11, 13, 19–23
    household uses of, 105
    solar heating of, 23–25
    specific heat of, 25–26
    storage of, 27, 113–14
    in thermosiphoning systems, 19–21
Water bags in solar heating (*see* Skytherm)
Water column, 29–30, *29* (*see also* Drumwall)
Water energy
    applications of, 104–5
    availability of, 102–4
    control of, 112–14
    conversion devices in, 105–8
    conversion factors in, 110–11
    sources of, 3
    storage of, 112–13
Water energy systems
    seasonal considerations in, 113
    siting of, 115
    sizes and efficiencies of, *109*
    speed control in, 112–13
    use of turbines in, 107–8
Water heater
    retrofit to solar, 51–52
    wood-burning, 100–1
Water-heating
    for hot water uses, 51–52
    primary vs. secondary systems of, 23–25
    for space heating, 27
Water-medium collectors, 9–11, 13, 37–41
Water pressure, 103–4
Water pumping
    in solar energy, 21–22, 38

    systems, 89–92
    from water energy, 104–5
    from wind energy, 89–92
Water rights, 115–16 (*see also* Building codes)
Water turbines, 107–8, 114
Waterwheels, 106–7, 114, 135, *135*
Weather, effect of solar usage, 34–35
Weatherability of solar collectors, 37, 41, 44
Weir, 109, 132–34, *133*
Wincharger, 65, 69, 80, *81*, 92
Wind
    effect on solar collectors, 37, 44
    types of, 61
Wind-electric machine (*see* Wind energy systems)
Wind energy
    characteristics of, 60–94
    conversion to electricity, 63–65
    power formula for, 62, 129
    for pumping water, 89–92
    source of, 3
Wind energy systems
    aeroturbine, 65–74 (*see also* Aeroturbine)
    building your own, 88–89
    construction of, 92–94
    power rating for, 92
    tax credit for, 94
    towers in, 86–89
    uses of, 83–86
Wind machines (*see* Wind energy systems; Aeroturbine)
Windmill, 60, 69, 70 (*see also* Turbine)
Windmilling (*see* Spoiling in windplants)
Window collector, 12, *12* (*see also* Air-medium collector)
Windplant
    rating of, 75
    spoiling in, 80
    stalling in, 81
    turntable, 66–67
    (*see also* Wind energy systems)
Windpower formula, 73–74, 129
Windsail, 67, 69–70, *69*, 91, 129
Windspeed
    in aeroturbine efficiency formula, 62
    average annual rating of, 63
    in cube law, 63
    estimates above ground, 129
    formula, 129
    rating, 74, 75
Wood-burning devices, 98–101
Woodcutting
    hardware, 101–2
    methods, 97
    permits for, 96
    safety, 102
Wood energy
    availability of, 95–97
    as a secondary system, 95
    uses of, 98
    yields, 97–98
Woodlots, 97
Wood, types of, 98

Yukon stove, 99, *99*

Zomeworks, 50

Michael Hackleman is a Renaissance man. While an electrician in communications and radar for the U.S. Navy, he became disillusioned with the state-of-the-art of nuclear waste disposal and began to study alternative energy. His formal education has been in drafting, electrical and mechanical engineering, electronics, mathematics and physics. He is the Research Director of Earthmind, a non-profit research organization. In 1973 Hackleman was credited with adapting Savonius rotors for the generation of electricity.

Hackleman, his wife Vanessa and son Brett live on a farm which utilizes many of the alternative energy systems described in his books. His latest project is building a Hybrid (wind-charged electric vehicle).